Synthetic Aesthetics

Synthetic Aesthetics

Investigating Synthetic Biology's Designs on Nature

Alexandra Daisy Ginsberg, Jane Calvert,
Pablo Schyfter, Alistair Elfick, and Drew Endy,
with additional contributors

The MIT Press   Cambridge, Massachusetts   London, England

MIT Press books may be purchased at special quantity discounts for business or sales promotional use. For information, please email special_sales@mitpress.mit.edu.

This book was set in Plantin and Gotham Rounded by the MIT Press. Printed and bound in the United States of America.

Library of Congress Cataloging-in-Publication Data
Synthetic aesthetics : investigating synthetic biology's designs on nature / by Alexandra Daisy Ginsberg ... [et al.].
p. ; cm.
Includes bibliographical references and index.
ISBN 978-0-262-01999-6 (hardcover : alk. paper)
I. Ginsberg, Alexandra Daisy.
[DNLM: 1. Synthetic Biology—trends. 2. Art. 3. Biotechnology—trends.
4. Esthetics. QT 36]
QP517.B56
610.284—dc23
2013013618

10 9 8 7 6 5 4 3 2 1

Contents

Acknowledgments   vii

Introduction: How Would You Design Nature?   ix

Part One   Designs on Nature

1   Synthetic Biology: What It Is and Why It Matters   3

    Alistair Elfick and Drew Endy

2   Countering the Engineering Mindset:
The Conflict of Art and Synthetic Biology   27

    Oron Catts and Ionat Zurr

3   Design as the Machines Come to Life   39

    Alexandra Daisy Ginsberg

Part Two   Nature versus Design

4   Nature Is Designed   73

    Drew Endy

5   There Is No Design in Nature   87

    Pablo Schyfter

6   Design Evolution   101

    Alexandra Daisy Ginsberg

Part Three   Synthetic Aesthetics

Logic

7   Bio Logic   143

    David Benjamin and Fernan Federici

8   Scale and Metaphor   155

    Pablo Schyfter

Process

9   Living among Living Things   169
    Will Carey, Wendell Lim,
    Adam Reineck, and Reid Williams

10 Constrained Creativity: An Engineer's Perspective   181
    Alistair Elfick

Time

11 The Biogenic Timestamp:
    Exploring the Rearrangement of Matter
    through Synthetic Biology and Art   195
    Oron Catts and Hideo Iwasaki

12 Time as Critique   205
    Jane Calvert

Structure

13 Synthetic Sound from Synthetic Biology   219
    Chris Chafe and Mariana Leguia

14 Abstraction and Representation   231
    Pablo Schyfter

Evolution

15 Living Machines   247
    Sheref Mansy and Sascha Pohflepp

16 Evolution or Design?   259
    Jane Calvert

Cultures and Context

17 The Inside-Out Body   271
    Christina Agapakis and Sissel Tolaas

18 Transgressing Biological Boundaries   283
    Alexandra Daisy Ginsberg

Epilogue: Reflections on Synthetic Aesthetics   295
Notes   303
About the Authors   319
Source Notes   327
Index   335

Acknowledgments

We would like to express our gratitude to all of those whose help was crucial to the project and this book. Neither would have been possible without the generous support of the Engineering and Physical Sciences Research Council (EPSRC) and the National Science Foundation (NSF) and their ongoing enthusiasm for our work. We would also like to acknowledge support from a Royal Academy of Engineering Ingenious Award, a Wellcome Trust Arts Award, the Economic and Social Research Council (ESRC) Innogen Centre, and the School of Engineering at the University of Edinburgh. Dozens of other institutions around the world have been vital to our work. We are grateful to the various workspaces and their members—laboratories and studios alike—that hosted the residents and supported this endeavor. We have also benefited from feedback from those who have been part of many different conversations, symposiums, lectures, and workshops associated with the project.

In writing this book, we have received thoughtful and useful comments from several colleagues. We are especially indebted to Christina Agapakis, Patrick Boyle, David Crowley, Emma Frow, and Sheref Mansy for their help in this capacity. We also wish to thank Kellenberger-White for creative direction, China Brooks-Basto for production assistance, Drusilla Calvert for indexing, as well as all of those who have contributed images to this book. We would particularly like to acknowledge all the administrative and support staff at the University of Edinburgh, Stanford University, and elsewhere who made such a big, international project happen.

Finally, we would like to thank the Synthetic Aesthetics residents and all the contributors to this book for their participation, their ideas, and their commitment to the project well beyond its official time frame. These individuals' energy and creativity made both the project and this book possible.

Introduction:
How Would You Design Nature?

Circuitry, toggle switches, gates, sensors, oscillators. This is the language of
component parts and manufacturing, of robots and computers and digital
logic. It is not the language of life and death, of protein tangles, evolution,
reproduction and decay, the everyday struggles of biological matter. Yet this
is now biology, albeit a new engineering approach to bioscience—the emerging field of synthetic biology.

The nineteenth century was shaped by the mechanization of the
Industrial Revolution; in the twentieth century, the silicon circuitry of an
Information Revolution restructured modern life. Now, some predict bio-
technology will be the foremost driver of change for the twenty-first century,

and synthetic biologists believe that their work will be integral to the success of this envisioned "Biotechnology Revolution" through the intentional design (or redesign) of biology.

Synthetic biology is a young field with growing global momentum, enticing engineers, biologists, chemists, physicists, and computer scientists to the laboratory bench to manipulate the stuff of life. These self-styled pioneers of biological engineering aspire to redesign existing organisms using engineering principles like standardization; some even seek to construct completely novel biological entities. The field's engineering vision leads to parallels being drawn with the early days of computer technology, as researchers reimagine bits of DNA code as programmable parts, analogous to the components of computer software and hardware (figure I.1). At the human scale, some synthetic biologists compare their culture to the garage innovators of the 1970s and 1980s who built the first personal computers and laid the foundations of a new industry. For synthetic biologists, biology could be just another material to engineer, its living machines driving twenty-first century progress.

What motivates this desire to make biology predictable and functional, to design biology rather than to understand it? Many synthetic biologists aspire to improve so-called genetic engineering. For these researchers, genetic engineering is less engineering than craft; it is an approach that can deliver unique products but not systematic tools and techniques. A genetic engineer may transfer the gene for an antifreeze protein from a fish into a tomato to make cold-resistant fruit, but the solution is only a one-off. Synthetic biologists instead hope to lay the foundations for a faster, more efficient, repeatable, and ultimately cheaper way to engineer living materials. Just as the standardization of the screw thread united individual manufacturers and users of nuts and bolts, and thereby helped drive the Industrial Revolution, this kind of bioengineering, it is hoped, will enable a Biotechnology Revolution. In short, synthetic biologists want to be reliably able to insert an antifreeze gene into any number of other organisms, including bacteria, with predictable results every time. Biology doesn't necessarily work in this way, but by applying engineering design principles—such as standardization—synthetic biologists seek to transform it (figure I.2). Future biological designers may even work far from the lab bench, dragging and dropping component parts using design software similar to that used by architects or programmers, expecting the same level of control over the materials they engineer.

This technical ambition is driven by dreams of plentiful, sustainable fuel, new manufacturing techniques, novel drugs and materials, and medical technologies (figure I.3). Through synthetic biology, living things could become both the operating system and the machine, in theory creating a technology

so versatile that it could be used to produce the food for a projected global population explosion and remediate the environmental damage wreaked by two centuries of industrial modernization.

This vision of a biology transformed into a medium and material for design is accompanied by grand rhetoric of a world-changing, world-saving green technology. Although such ambitions are admirable in their scope, they raise many questions. What is the potential for unintentional, or even intentional, damage caused by biotechnologies? How are we to manage the ownership of life's materials? These issues have been and continue to be much scrutinized by bioethicists, social scientists, and policy makers. But

Figure I.1
Revolution or evolution? A film of genetically modified, light-sensitive bacteria displays the classic computer program message. These "*E. coloroid*" bacteria were engineered by undergraduates from the 2004 University of Texas, Austin and University of California at San Francisco iGEM team.

Figure I.2
Bacteria have become the
workhorses of synthetic biology.
Here, biologist Fernan Federici
"labels" *Bacillus subtilis* to
fluoresce, tracking their
self-organized pattern formation,
as seen in an optical microscope
at ×1000 magnification.

Figure I.3
Synthetic biology is described
as a transformational technology.
In 2010, synthetic biology
students at the University of
Cambridge engineered these
Escherichia coli bacteria—dubbed
"*E. glowli*"—to bioluminesce
extra-brightly.

synthetic biology also presents complex new issues that are not often discussed, such as the path of the technology's development, the direction we want it to take, and how the aims of synthetic biologists align (or fail to do so) with those of the technology's potential users: us all.

Despite being a young field, countless reports have been written on synthetic biology and the social, ethical, and legal issues it raises.[1] Up until now, much of this discussion, both of sustainable futures and risk, has been speculation. But now we are at a point in time in which synthetic biology is becoming increasingly mainstream and is receiving growing financial support across the globe. At the time of writing, the largest public funders of the field are the Chinese, U.S., and U.K. governments. As synthetic biology develops, its practitioners are beginning to orient their work toward industrialization, slotting into existing and accepted ways of manufacturing. There is a danger that synthetic biology will become myopic and monolithic, following the well-trodden path of industrialization, including first generation industrial biotechnology. Synthetic biology may simply become a way of pumping out more of what we already have—such as fuels or plastics—using biological rather than non-biological processes. This new technology could be used to give a green gloss to harmful practices like inefficient production, excessive consumption, and toxic waste—the problematic aspects of "successful" industrialization (figure I.4).

Alternative visions of synthetic biological consumer products range from the mundane or frivolous (like probiotics and diet pills) to the imaginative and challenging (such as plants engineered for pleasure or living building materials). Could synthetic biology perhaps change our lives in these unexpected ways? The promise of the technology may well be no more than hype, yet these discussions demand society's attention and participation. They should not be limited to a select few with a controlling stake in the technology.

This book about synthetic biology is unusual in that it presents an ongoing dialogue between synthetic biologists, artists, designers, and social scientists, all with very different views on this emerging technology. We draw on a diversity of perspectives and projects to explore and challenge the understanding of design in synthetic biology. Our aim is to provoke discussion about what place—if any—design should have in our relationship to living things. What does design in synthetic biology really mean and what might it involve? What responsibilities does designing biology carry, and what consequences could it have? This focus on design allows us to question, challenge, and reconsider the assumptions made about the future of this developing technology, one normally rendered through contradicting visions of utopian green salvation or dystopian bio-apocalypse. We are

seeking ways to understand better the scientific, technological, ethical, philosophical, political, and social dimensions of synthetic biology using art and design to identify new areas for enquiry.

Instead of finding solutions to predefined problems, we propose that we should be challenging the questions that are being asked. We see many reasons for advancing alternative perspectives on synthetic biology, as it is in the process of being developed. First, there are technical arguments about biology itself. Rather than treat living nature as just another material for engineering, synthetic biology may benefit from engaging with its unique properties, which, though complex and unpredictable, might suggest new approaches and perspectives to using life as a raw material. Second, synthetic biology is often promoted as a sustainable solution to our manufacturing and energy woes, but there is a paradox in this reasoning. Industrialization and design are oriented toward growth, not equilibrium

Figure I.4
Will synthetic biology simply feed into existing systems of use, consumption, and waste or could we design more from it? Photographer Chris Jordan documents today's detritus in "Intolerable Beauty: Portraits of American Mass Consumption, Crushed Cars #2, Tacoma 2004."

and sustainability. Biology grows within the balance of ecosystems, but can commercial synthetic biology be a sustainable, renewable technology on a planet with finite resources (figure I.5)? There may be alternative strategies to explicit industrialization that could better address the problems that synthetic biology purports to solve; approaches that are novel, imaginative, and more suitable for designable biology. Engineering biology appropriately could help us address profound problems in the logic of production and consumption that underpin design and engineering today. But it is clearly not the only way to address these challenges. It is important to ask when and whether we should be turning to synthetic biology, rather than to other technical, social, or political solutions. Asking disruptive questions like this may not be comfortable, but it can be productive, making things visible that otherwise would not be so. Our aim is not celebration, but exploration and interrogation of the expectations and limitations of synthetic biology.

Figure I.5
Massive algae bloom in 2011 at Qingdao Beach, China, triggered by water pollution.

The Synthetic Aesthetics project team comprises two synthetic biologists with engineering backgrounds, two social scientists, and a designer/artist. As part of the project, we have all been forced to engage with unfamiliar perspectives and to challenge our disciplinary assumptions, finding new ways of working that have taken us beyond what we could have done in our separate disciplines. Synthetic Aesthetics is not specifically an engineering or science, social science, or art or design project but it draws on all these areas. Although our focus is synthetic biology, we are investigating the values and assumptions that could affect us all, beyond those actively working in the field.

We want to generate discussion about synthetic biology, its aims, and its potential implications, using art, design, and social science to transcend the narrow and one-dimensional way in which it is starting to be framed in order to stimulate more improbable and creative thinking. This requires not merely considering our needs today, but enabling novel paths that will allow biology, technology, and society to interact in new ways in the future. We hope to increase the range of possibilities and future trajectories for technological development and to promote better outcomes—with the recognition that what is meant by "better" is something that must be actively and continually debated.

Origins

The unusual origins of the Synthetic Aesthetics project help explain the unconventional nature of the research. The project was conceived, developed, and funded over just five days as part of a workshop held outside Washington, D.C., in March, 2009. The "IDEAS Factory Sandpit on New Directions in Synthetic Biology" was organized and funded by the Engineering and Physical Science Research Council (EPSRC) of the United Kingdom and the National Science Foundation (NSF) of the United States. Leading academics in synthetic biology from across the United States and the United Kingdom came together to develop grant proposals from scratch. "Sandpits" are not a normal research funding mechanism; they are an unconventional method used to foster innovative interdisciplinary research proposals in a very short time period, in contrast with the typical lengthy grant-writing process accompanied by the conservatism of peer review common in science today. Sandpits are intense: Participants subject each other's proposals to "real-time" peer review, successful projects are funded by the close of the workshop, and new collaborations are cemented.

An ideal sandpit project is meant to be multidisciplinary, transformative, novel, and innovative, and participants are encouraged to put forward risky and adventurous proposals encroaching on new territory. "Synthetic aesthetics" started as an anonymous phrase written on a Post-It note and stuck to a wall. It captured the interest of the three of us who attended the sandpit (Jane Calvert, Alistair Elfick, and Drew Endy). We speculated about what would happen if we initiated collaborations between synthetic biologists and creative communities that allowed both groups to imagine their work in new ways. We took seriously the organizers' encouragement to "think outside the box" by performing a dance based on the myth of the Golem to present some of our early ideas. We had discussions about the sublime. We were challenged by our colleagues to define beauty and defend frivolity. What resulted from this strange and intense experience was a project that was completely unexpected. When we wrote our original proposal at the sandpit, one of our hypotheses was "we will be surprised." This has proved to be a good working hypothesis.

During the sandpit and since, some people have assumed that our aim is outreach: a public relations activity on behalf of synthetic biology to beautify, package, sanitize, and better communicate the science. We reject and actively resist such a framing. Our project has been an exploratory investigation of the intersection between art, design, and synthetic biology, encouraging dialogue and dissent. Creating a space for critique continues to be a guiding principle of Synthetic Aesthetics.

The Synthetic Aesthetics Residencies

The core of the project has been the curation of paired residencies between six artists and designers and six synthetic biologists working in Europe, Asia, Australia, South America, and the United States. Their shared expertise extends across the spectrum of synthetic biology, from plant science to protocell research, and encompasses a diversity of approaches to art and design, including architecture, music, smell design, bio art, and product design.

We received hundreds of applications from designers, architects, writers, dancers, painters, artists, chemists, biologists, computer scientists, and engineers who wanted to participate in the project. We intentionally selected artists and designers whose work was not primarily concerned with visualization—translating science into images or objects. Instead, we chose those who were interested in directly engaging with the subject matter of synthetic biology. The Synthetic Aesthetics team matched up the pairs (except for Oron Catts and Hideo Iwasaki, who applied together). We sought out themes that linked the residents' work—some almost imperceptible at first—and that we hoped would generate unexpected insights (figures I.6 and I.7).

Tasked with investigating design and synthetic biology, the Synthetic Aesthetics residents had explicit freedom to take their work in any direction they chose. Art/science projects often involve artists visiting—even working—in labs, but it is unusual to extract scientists from their work environment and

Figure I.6
Workspace of a Synthetic Aesthetics resident: Christina Agapakis's lab bench in the Silver Lab at Harvard University Medical School.

Figure I.7
Workspace of a Synthetic Aesthetics resident: Sissel Tolaas's smell-molecule studio in Berlin.

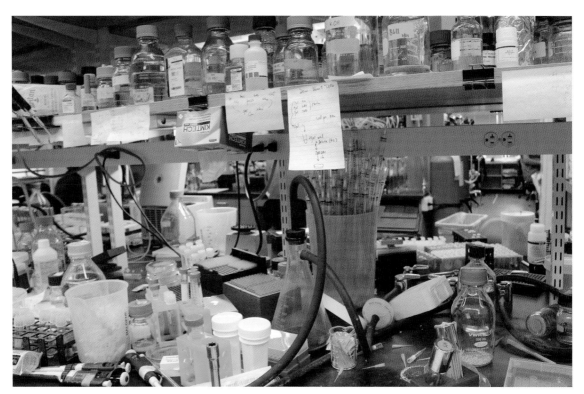

put them in a studio as part of these collaborations. For our program, each residency began with an intensive two-week period in one partner's laboratory and then continued by moving to the artist/designer's studio for two weeks. Both stages were documented by the project team. These two-way exchanges were intended to encourage reciprocal collaborations that could contribute to both partners' practice. At the time of writing, two years after the residencies began, all our residents continue to collaborate. Their research together is described in their contributions to Part Three of this book.

We were encouraged that a large number of scientists and engineers wanted to participate in a project beyond the remit of their "normal" responsibilities. It may be that aspects of synthetic biology make it well suited to this type of unconventional initiative. It is already an extremely interdisciplinary field, so it may not be as much of a stretch to involve artists, designers, and social scientists as it might be in a more established, traditional discipline. In addition, some synthetic biologists explicitly aim to "make biology easier to engineer" and to make the science more accessible to the outsider. But perhaps the most important reason why synthetic biologists are open to these kinds of collaborations is because they are part of a new field that has not yet stabilized. The relationships between science, engineering, and society are still being created and negotiated, providing opportunities for interdisciplinary investigation, experimentation, and debate. Through the project, we wanted to increase the range of people who have a voice in the future of synthetic biology and who can contribute to decisions about the directions it may take. Our long-term aim is to enable new groups of practitioners, thinkers, and critics to engage with developments in synthetic biology and to broaden the conversation about how we should best make use of our abilities to manipulate the natural world. But as the field becomes more established, will initiatives like Synthetic Aesthetics be harder to instigate?

Nature, Biology, and Design

Part One of this book introduces our two key areas of interest: synthetic biology and design. In chapter 1, Alistair Elfick and Drew Endy, both engineers by training, introduce synthetic biology and describe their vision of using biology as a material for design by harnessing its unique properties. In direct contrast to the engineers, in chapter 2 artists Oron Catts and Ionat Zurr express their concern that we are moving into a future dominated by a single engineering paradigm. They argue that the "engineering mindset," which has developed over the past century, threatens to monopolize life. One way of drawing attention to alternative frames of thought, they maintain, is to open up the tools and spaces of biotechnology to other disciplines, including art. In chapter 3, the designer and artist Alexandra Daisy Ginsberg investigates the "design mindset" and the view we have of ourselves and the things that we make as somehow emancipated, or separate, from nature. Synthetic biology is modeling itself on existing design disciplines, yet design itself is part of the production system that is in crisis, preventing sustainable, ethical, and imaginative innovation. She argues that there is an opportunity to reinvent what we understand as design as we reinvent biology.

Part Two explores nature and design from three different perspectives. In chapter 4, Drew Endy maintains that evolution "designs" and "problem-solves," producing solutions that provide inspiration for synthetic biology and giving weight to the idea of nature as a materials library. In chapter 5, social scientist Pablo Schyfter puts forward the contrary view that there is no design in nature. He argues that because values are central to design, design is something that only people do. Importantly, in synthetic biology, design introduces the question of values into a domain previously independent of them. Alexandra Daisy Ginsberg in chapter 6 shows that it is crucial to recognize that synthetic biology's designs will inevitably be part of a much larger social, political, and economic context. Synthetic biology makes use of engineering design principles, but she suggests that we should also develop human-scale design principles in order to understand better what good or bad design might mean for biology.

Our six case studies, the Synthetic Aesthetics residencies, are presented in Part Three, each exploring the idea of design in synthetic biology in different ways. In our responses to each residency, we expand on the issues and themes each pair identified through their joint projects. The aim is not to present finished work, but to offer new models of interaction and discussion and suggest new ways of working.

The residencies raise a host of challenging questions. There are questions about scale and form, such as: can design tools move across scales, from the biological scale of micrometers to the architectural scale of meters (chapter 7; David Benjamin and Fernan Federici)? How can synthetic biological designs be represented in sound and music (chapter 13; Chris Chafe and Mariana Leguia)? And what are the limitations of the abstraction that is required for design in synthetic biology (chapter 14; Pablo Schyfter)?

There are also questions about process: how to enable cross-disciplinary collaboration between synthetic biologists and designers (chapter 8; Pablo Schyfter)? Can ideas of "design thinking" from commercial product design be successfully applied to science education and research (chapter 9; Will Carey, Wendell Lim, Adam Reineck, and Reid Williams)? Has engineering, to its detriment, delegated creativity to designers, and if so, can it be recaptured through synthetic biology (chapter 10; Alistair Elfick)? Then there are questions about the specificities of designing with a biological substrate.

Are evolution and design totally separate activities or is evolution one of the tools that can be used in design (chapter 16; Jane Calvert)? As we potentially return to living machines after a period of inanimate mechanization, might the living machines we design have their own agenda (chapter 15; Sheref Mansy and Sascha Pohflepp)? Finally, there are questions that provoke us to think about synthetic biology in new ways. Do radically different temporal perspectives on synthetic biology (such as geological time) challenge us to rethink our biological designs (chapter 11; Oron Catts and Hideo Iwasaki)? What is the difference between art and design in their critical engagements with synthetic biology (chapter 12; Jane Calvert)? How might synthetic biological designs encourage us to reimagine our relationship to food and bacteria (chapter 17; Christina Agapakis and Sissel Tolaas)? And how can our own bodies give us insights into synthetic biology and design (chapter 18; Alexandra Daisy Ginsberg)?

None of these questions have easy answers. They are the starting point for the challenging and stimulating conversation about synthetic biology and design that we hope to initiate. We want to provoke readers and inspire new questions. We are prescriptive only about the importance of thoughtfully and critically exploring synthetic biology and its place in our world.

Part One
Designs on Nature

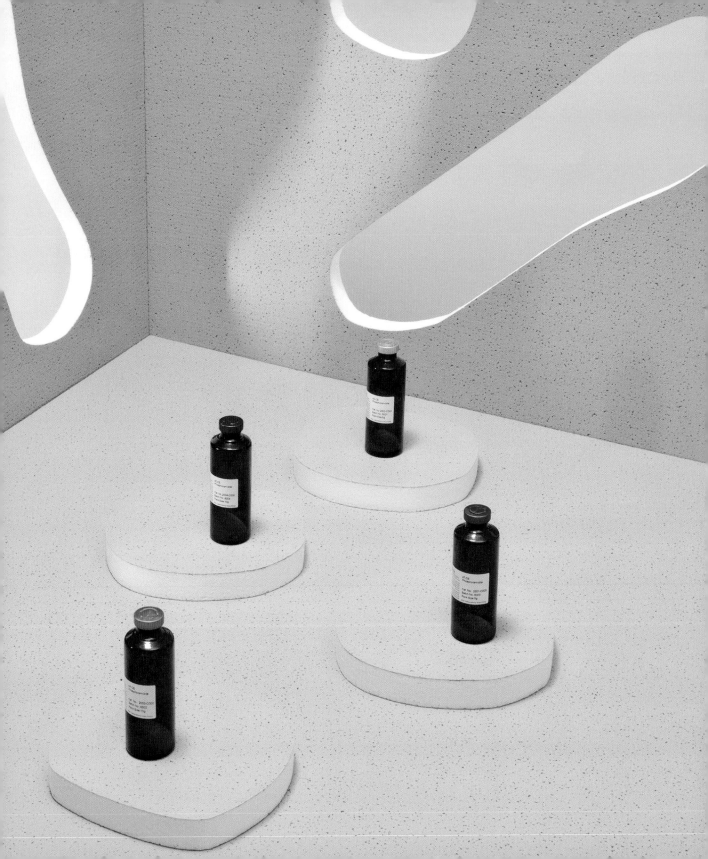

1.
Synthetic Biology:
What It Is and Why It Matters

Alistair Elfick and Drew Endy

Menlo Park is a town near San Francisco, California. In 2010, the then estimated 32,026 residents of Menlo Park collectively threw away approximately 7 million kilograms of newly manufactured nanotechnology products, or roughly 220 kilograms for every man, woman, and child. These products were made using a sophisticated distributed manufacturing process in which each plot of land received necessary raw materials, such as water, from a supply system established over the past 100 years. On each plot, natural resources including atmospheric carbon dioxide were carefully incorporated into manufacturing systems with atomic precision, producing macroscopic structures capable of growth, assembly, healing, and reproduction. The

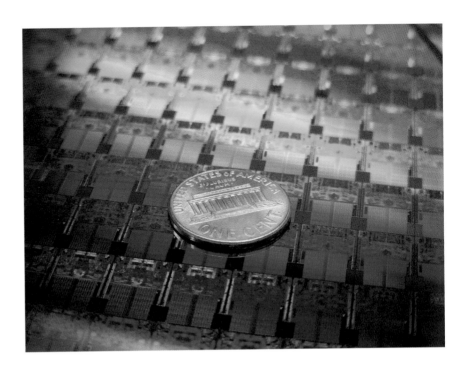

Figure 1.1
A silicon wafer containing
hundreds of chips with billions
of manufactured features. Its
dimensions are such that 1,000
could be lined up across the
width of a hair. The penny shown
covers an area equivalent to four
computer chips.

entire process was powered by solar energy–harvesting systems that were themselves manufactured in real time, as needed. Nobody thought too much about these details, except for the planners and workers responsible for hauling everything away. Similar scenarios play out every year in thousands of cities worldwide. Fortunately, and in contrast to man-made nanotechnology products, the plant-grown products from Menlo Park will rot.

Meanwhile, in companies headquartered nearby such as Intel Corporation and Apple Incorporated, engineers worked to design next-generation microprocessors needed to make the computers of tomorrow. In 2010, these and other organizations manufactured and shipped roughly 400 million computer chips across the globe. A contemporary computer chip is made from sand that is purified and polished into a silicon wafer and then doped with boron, phosphorus, or arsenic atoms in order to create semiconducting material as seen in figure 1.1. The exact locations of dopant atoms are not controlled precisely but are instead distributed throughout a silicon layer randomly, not unlike chocolate chips scattered within a cookie. Nevertheless, the chip manufacturing process requires expensive, polluting, and energy-intensive centralized facilities that reliably implement "top-down" control of pattern formation on the silicon wafers. Each newly produced chip, once coated with protective materials, weighs about a single gram. Thus, in 2010, the physical matter composing the world's annual supply of computer chips weighed less than 400 metric tons, or roughly equivalent to the weight of just

400 family-size cars. Many well-paid experts thought carefully about every detail of their making, from raw materials sourcing to design engineering, and from manufacturing to marketing and sales of the finished consumer electronics. Yet, all these silicon chips amount to just 6% of the manufacturing capacity to be found within Menlo Park's lawns and gardens.

Instead of producing precision materials that were immediately recycled, could the citizens of Menlo Park ever expect that their town's biological manufacturing capacity might eventually produce the world's annual supply of computing hardware? To begin to explore this question, we need to consider more directly the nature of Menlo Park's apparent surplus manufacturing capacity. For example, the Endy front yard is blessed with a mature Pacific Silver Fir tree. Each summer many hundreds of pinecones take shape, growing up from branches throughout the tree. Each cone starts as a mostly amorphous expanding blob of green sap, quickly growing to be 10–15 centimeters tall and 6–9 centimeters wide. Over a period of weeks, the structure of the cone takes on precise repeated patterns and, not unlike that seen in figure 1.2, matures ready to release its payload. Each cone is a close replica of its neighbors on the tree and, for that matter, on any other Pacific Silver Fir. Within the mature cone there are hundreds of seeds, each possessing the genetic information necessary to instruct the programmed self-assembly of another instance of a Pacific Silver Fir, complete with its own cones.

Figure 1.2
Three stages of the self-assembly and maturation of a "seed dispersion subcomponent"—a pinecone. The 1-gram seed that the pinecone releases holds the complete information required to execute the programmed assembly of an entire tree and its subsequent ongoing manufacturing capacities.

Biology's remarkable ability to execute precision growth of highly complex structures dwarfs ours. Myriad opportunities exist in harnessing the power of biology for the programmed manufacture of human artifacts that possess precise functions. The seemingly wild speculation that a pinecone could instead be programmed to differentiate into a stack of microprocessors is worthy of exploration. What gives biology the ability to achieve these feats?

Consider that the shell of an abalone, a large marine snail, is 98% calcium carbonate and only 2% protein but is 3,000 times stronger than a geologically constructed equivalent material, such as chalk. The key to this extraordinary strength is in the nanoscale precision of its manufacture with highly regular calcium carbonate tiles bound by protein glue. Within the tissues of the snail, molecular machines are made that possess the ability to capture a given atom and place it exactly where needed within a growing tile. If we could borrow the manufacturing skills of the abalone and combine these with the light-harvesting properties of plants, maybe we would produce an organism that could make new super-strong building materials with nothing more than sunlight and seawater. What then if we could move biology's capacity forward? Having learned its tricks, can we teach it new ones? Can we ask these molecular machines not to shuttle calcium about, but rather silicon and dopants so we can build pinecone processors? It seems like a wild piece of science fiction, but maybe, just maybe, it is possible. Certainly, the incredible silicon oxide manufacture discussed in chapter 4 suggests that this isn't as outlandish as it sounds. Phosphorus is a key constituent of many biomolecules, including DNA. Boron is known to be essential in minute amounts for healthy bone growth and plant embryogenesis. Arsenic is toxic to many organisms, prompting some of them to evolve pump machinery to excrete it while others use it as an electron acceptor. So biology can handle all these silicon dopants in one way or another.

The molecular machines described above are one of a huge family of such nanoscale workers. The members of this family possess countless different specialties, like being able to splice molecules together, cut parts off molecules, bind specific atoms, pump ions, and untwist molecules; in fact, we actually only know the functions of a small number of this fraternity. In addition to precision manufacturing, these molecular machines can assist in the synthesis of complex molecules, many of which humankind has been using for a long time. For millennia, we have been sourcing molecules from organisms for things like flavors, fragrances, vitamins, and drugs. With synthetic biology, we can harness the machinery to make those molecules such that we don't need to go destroying the planet's ecosystems to harvest them.

One final capability that we, as engineers, greatly admire in biology is the exquisite sensitivity and specificity that organisms can achieve in molecular recognition, or sensing. This is best exemplified when thinking of the sniffer dog's ability to discern the odor of tiny quantities of narcotic drugs among the background bouquet of an airport customs check. This talent for sensing is finding ready use in many applications, from the sensing of arsenic contamination in drinking water,[1] a considerable problem on the Asian subcontinent, to the monitoring of environmental pollution.

As a consequence of about 4 billion years of evolution, nature has become very capable in its use of raw materials and energy. Ecosystems are underpinned by organisms that have the ability to meet all their needs from the environment—the carbon, hydrogen, nitrogen, oxygen, phosphorus, and sulfur that compose 99% of their tissues and the trace elements no less vital for their health such as iron, copper, cobalt, zinc, and manganese. These atoms are then organized together using the energy at their disposal, be it chemical, solar, or geothermal. Once captured, this energy and raw material can then be transmitted, often with few losses, throughout the ecosystem up the food chain. The progression of biology from a mere set of chemical reactions driven by some extrinsic energy to become *life*, in all its diversity, is truly wonderful. Or to put it succinctly—life is simply breathtaking.

The motivations to harness the powers of biology sympathetically are then evident—it would be incredible to partner with biology to deliver a whole new engin eering paradigm for the twenty-first century. A biotechnology that is capable of making much of what our civilization needs that is also responsible, equitable, and founded on the invention of a sustaining relationship between humanity and nature.

A History of Material Appropriation

Humankind has an extended history of adopting new materials with which to make things. Our pre-human relatives began to adopt found objects as tools some 3.4 million years ago, using fist-sized stones for tasks such as processing food. The shaping of those found objects to form tools, such as hand axes, and the adoption of these tools in everyday use is believed to have been highly influential in supporting the ascent of humans. By 1.8 million years ago, tool use had directed the evolution of our ancestors, their use of the inanimate enabling the genetic adaptations needed to move from tree-dwellers to the meat-eating, ground-dwelling *Homo erectus*.

Around 6000 BCE, the ability to smelt metal was discovered: first copper, then bronze, then iron, and then steel. With each new metal adopted, the scope of possible products changed. Ways of making altered. New metal-forming processes were developed: casting, forging, turning, milling, drilling, and so on. These events presaged the dawn of the Industrial Revolution, a historical step-change when making became manufacturing. Products ceased to

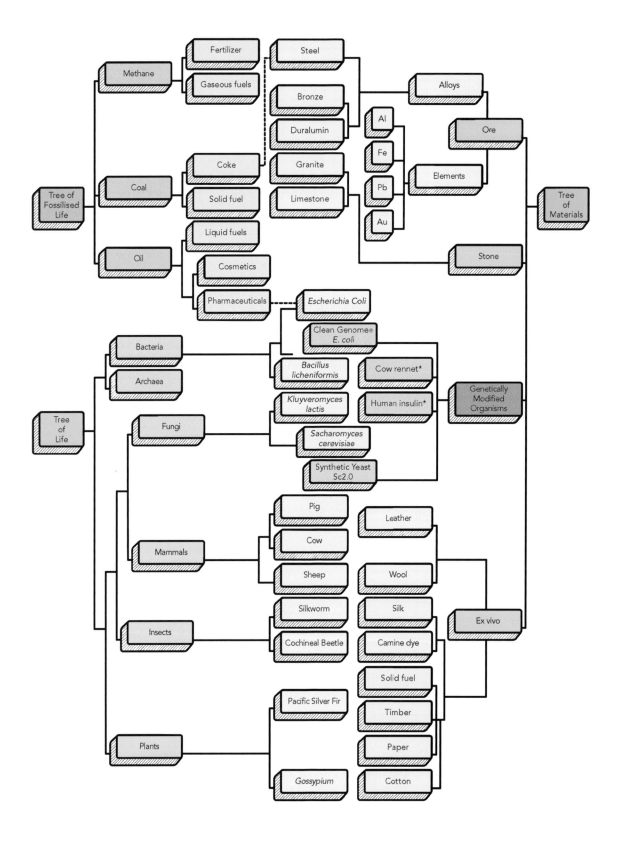

be handmade, instead becoming machine-made: highly repeatable instances of an artifact, produced rapidly. Initially these were premium products able to command higher prices than their variable, handmade precursors; only latterly has mass production become associated with reduced cost. But the tangible impact of adopting new ways to make things did not stop at cheaper, better, faster production. The types of products being made changed, existing products diversified, and completely new product types, such as automobiles, were created.

The palette of materials from which it is possible to make things is now very diverse. The evolution of these materials as the substrates for twenty-first century making has happened over a range of timescales; some materials, such as copper, have been at our disposal for millennia and others, such as polymers, for only a few decades. Drawing an analogy to the tree of life, we can conceive of a tree of materials that illustrates the evolution of objects made from all the various types of materials. If synthetic biology proposes that living organisms can become a substrate for engineering and expanded making, a new material in our palette, how would the tree of materials begin to intersect with the tree of life? Polymers are made from (mineral) oil, which in turn is fossilized biology, so where on our two trees would the products made from polymers be found? We present in figure 1.3 a possible realization of this notion that attempts to consider the degree to which a raw material requires processing before becoming a substrate for making. Today, we are at a nascent stage in the process of entanglement in the branches in our trees, way up in the canopy; the hybridization of the tree of materials with a new branch in the tree of life, *The Synthetic Kingdom*, as proposed by Alexandra Daisy Ginsberg (see chapter 3).

How Do You Make Biology Engineerable?

Marveling at the capacities of biology to construct objects and boldly stating an aspiration to harness that capability is all well and good. But how do you actually go about realizing this ambition? Is it even possible? To answer this requires a brief introduction to a molecule well known by its acronym: DNA.

In 1953, James Watson and Francis Crick published their Nobel-winning work solving the structure of deoxyribonucleic acid (DNA)[2] based on data produced by Rosalind Franklin. The new science of molecular biology was heralded as revealing the physicochemical basis of the fundamental processes of life and heredity. DNA was identified as the universal information storage molecule across the whole of nature. There followed two decades of exceptional scientific discoveries deciphering the process through which proteins are synthesized within cells. This led to Crick's 1970[3] reiteration of "the

Figure 1.3
Conceptual, Linnean-style, *tree of life*, in green, adjacent to a *tree of fossilized life*, in orange, hybridized with an opposing *tree of materials*, in blue, with an epiphyte of *made-life*, in pink. The bottom branch on the *tree of materials* are substances harvested directly from nature. The upper, inorganic, branches of the *tree of materials*, often still bear some relation to the biological; limestone has a place on both the *tree of fossilized life* and the *tree of materials*.

central dogma of molecular biology," first postulated in 1958, that information, stored in the form of DNA, could flow via RNA to instruct a protein's construction and hence an organism's behavior. This view, the notion of a linear connection between the information held in a segment of DNA, called a gene, and a behavioral (phenotypic) outcome, still prevails. The antithesis, that an error in a gene may cause a malfunction or disease, is also much deployed shorthand.

Consider a conceptual factory making cars. You can easily imagine that there exists a single repository of information, some kind of secure database, which holds all the instructions about how to manufacture every single subcomponent and their subsequent assembly into the finished product. It is more than just a set of blueprints; it is a complete instruction source that captures all the detail needed to read and implement those blueprints. From this single data repository, subroutines can be published that go out to direct the production line; a subroutine for assembling the suspension, for example. In this way, the workers are not overwhelmed by having to find the relevant pages from within the whole instruction tome. The workers become specialized: Subsets of the workers become skilled in conducting certain tasks. It is also much easier to print the multiple sets of subroutine instructions needed for your workforce. The result is an efficient factory with skilled workers making quality products.

A cell may be thought of as a factory. How then does a cell manage the information it needs to live? Each cell has a single copy of its genome, a complete information storage repository fashioned from DNA. Some argue that the genome is more than just a set of blueprints; it is a complete instruction source that captures all the detail needed to read and implement those blueprints. From the master copy of DNA instructions, a shorter string of information may be written onto a new molecule in a process called *transcription*. The transcribed information is now contained in ribonucleic acid, or RNA, which has a slightly different structure than DNA. Many copies of RNA can be written from a single section of DNA allowing many subroutine instructions to go out to the "workers" of the cell, incredible molecular machines called ribosomes. The ribosomes are able to read the RNA command statements and use the information they carry to make protein molecules; this final step is called *translation*. A final important process for the information-storage molecule DNA is its copying in a process called *replication*. This allows for a complete copy of the DNA to be passed on to its daughter when a cell divides.

The subroutine information stored in the DNA is held in a unit called a gene. These are sections of the genome that encode a discrete "heredity element," which can exert an effect on the organism via the RNA or protein it produces. Each gene has a set of common features, though the details of their operation vary depending on whether they are from bacteria, animals, or plants. The important features can be summarized as follows:

Transcriptional promoter Each gene possesses a promoter that acts like a faucet, turning on and off the reading of a gene. The "strength" of the promoter helps to regulate the amount of gene product made by a cell. Promoters, and in particular conditional promoters where the promoter can be controlled, say turned on or off, are very important for giving the organism the ability to respond to environmental cues.

Transcriptional terminator This is a section in the DNA where its structure releases the transcriptional machinery. Termination can happen either as the structure of the newly formed RNA pulls itself out of the machinery or through the action of another molecule making the machinery stall. Transcription doesn't always continue to the terminator; the cells can further regulate transcription by halting it early, producing an RNA molecule that doesn't work (i.e., doesn't contain a complete copy of the information needed to instruct a ribosome how to make a specific protein).

Coding region This is the stretch of DNA that will form the section of RNA that is subsequently translated into the protein gene product.

Translation start and stop codons Specific regions in the RNA form sites for the ribosome to attach and release. For initiation, the start codon forms the focus for the assembly of a very complex machine formed from many molecules. This complexity is necessary because of the "punctuation-free" nature of the genetic code; the position of the start codon determines the way the information in the RNA is interpreted to make a protein. If this were to initiate in the wrong position, it would result in the complete distortion of the message and a useless gene product. For termination of translation, the stop codon is recognized by a release factor that causes the ribosome to fall off the RNA, indicating that the protein is complete.

If we were able to learn how to read, understand, and write DNA as a programming language, we ought to be able to generate desired function in an organism of our choice. In this case, the organism could be thought of as a chassis, an entity that provides the energy and infrastructure to execute the code we have written. Much of the behavior of the organism would be unchanged; it would still be able to grow, reproduce, and so on, as before. Fortunately, in DNA, nature has implemented a programming language that is highly modular. Common motifs are re-used such that we can adopt their function wholesale. This innate modularity lends itself to the notion

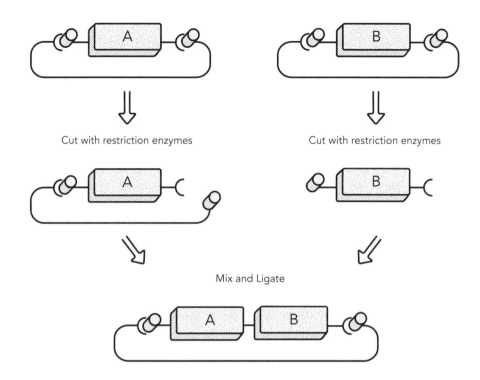

Cut with restriction enzymes

Cut with restriction enzymes

Mix and Ligate

of programming at this level. We can write genetic circuits, short programs, composed from the modular DNA parts that we are able to source from across nature. The simple but powerful notion of having an agreed method of connecting together the DNA parts means we can share each other's efforts. This standard construction for DNA parts, as can be seen in figure 1.4, has become known as the BioBrick standard, invented by Tom Knight. When stored in a library that is freely accessible to the community (e.g., the "Registry of Standard Biological Parts"), the power of the BioBrick standard is enhanced. The functional performance of all the DNA parts must be measured and recorded in a format that allows the data to be easily accessible to anyone wanting to use a given part. Datasheets can further facilitate this role, presenting information on agreed performance tests, like "what this device does," "how the device turns on," "how the device switches off" and so on.[4]

In wondering about an "engineering" approach to the adoption of biology as a material for making, reference is often made to the Industrial Revolution, as we have earlier. It is used to exemplify the benefits that can be found in working cooperatively and sharing information. Standards have been forged during the emergence of all modern technologies. A standard is really just a decided form of behavior—a common unit for measuring temperature, a shared way of drawing designs, an agreed method to measure the strength of a material. Standards help communities to work together by enabling the

Figure 1.4
BioBrick parts—standardized biological parts made from DNA— can be assembled using plasmids to make biological "circuits."

exchange of hardware, or knowledge, among a community of practitioners, facilitating the interoperability of machines and humans and driving a subsequent move to mechanization and mass manufacture. By adopting a standard composition from BioBrick parts, it is possible to share those parts freely among the community. For example, just by standardizing how students might coordinate the assembly of designer DNA fragments, Tom Knight's original BioBrick assembly standard allowed thousands of students to participate in a global genetic engineering "Olympics" known as the International Genetically Engineered Machine Competition, or iGEM.

Standards predate the profession of engineering and are thus not the sole preserve of the engineer. For example, we all use measurement standards in our everyday lives without giving their importance too much regard; each time you go to the grocery store, you trust that the weight of produce that you purchase is accurate. Standards for weights and lengths emerged with commerce and were reinforced by the shift toward a market economy based on value being abstracted into monetary instruments. Laws and regulations enforced societal compliance with the system of abstraction and standardization.

For a particular protocol or process to become a standard requires that it is accepted by a substantial community of practitioners. This process can be driven in a number of ways. Since the mid-twentieth century, adoption of standards across traditional engineering disciplines has sometimes been facilitated or codified by the creation of standard-setting bodies such as the Internet Engineering Task Force (IETF) or the International Organization for Standardization (ISO). Prior to this, standards were taken up in a more *ad hoc* manner. For example, the measurement of the power output of an engine is still widely given in horsepower, a historical anomaly. In the late eighteenth century, Scottish engineer James Watt had a problem. To make money from his efficient new steam engine, he had come up with the idea of taking royalties of one third of the coal saved when replacing earlier steam engine designs. The catch was if you replaced horses with the steam engine, there was no saving in coal; a loophole no Scot could tolerate! Thus, in 1782, Watt conducted an experiment in which he measured the power output of a "brewery horse" (a number of which could be found in Scotland—both breweries and horses that is). The measured figure gave 1 horsepower equivalent to about 10 men-power at 32,400 foot-pounds per minute, later simplified to 33,000 ft-lb/min[5] (see figure 1.5). Ever the canny Scot, Watt had measured the peak performance of the horse, and as such his engines would always overdeliver on their promised horse-replacing capabilities. This social engineering ensured that his definition of horsepower became widely accepted and that his engines quite literally drove the Industrial Revolution.

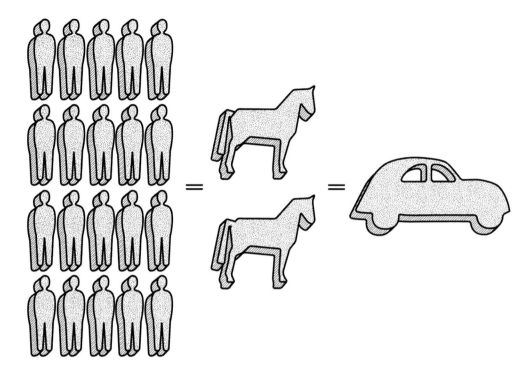

Figure 1.5
One horsepower is equivalent
to the power output that can
be sustained by about 10 men.
Equivalent to 746 watts of power,
it is the effort required to raise a
load of 76 kilograms by 1 meter
in 1 second. Internal combustion
engines range from a few to
many hundreds of horsepower.
The Citroën 2CV (in French:
deux chevaux, or "two horses")
illustrated here was a post-World
War II, Bauhaus-inspired economy
car designed to mobilize the
French proletariat.

Meanwhile, standardization of measurement in biology is a challenging process. An example of this would be the characterization of one of the simplest DNA elements, the transcriptional promoter. Promoters are essential to synthetic biologists who seek to program useful functions within cells, and hence good methods to coordinate the measurement of promoter activities would be a very important foundational technology. Instead, promoters are conventionally quantified by observing the amount of whatever product is made from the gene next to the promoter, but this is not a robust enough measure. The amount of gene product can be affected by a huge range of other things, including the specific gene itself. Introducing a "yardstick" is a good way to compensate for the contextual influences on a cell—promoter strength can be measured relative to a reference promoter.[6] Two identical populations of cells in the same growth environment are needed; in one you measure gene product driven by the reference promoter, and in the other you measure gene product driven by the promoter X. The strength of promoter X can then be expressed in "relative promoter units" (RPUs). A measure of say, 1.45 RPU would indicate that promoter X drives gene production 1.45 times more strongly than the reference promoter.

Thinking of a promoter as a faucet is a form of abstraction. By encapsulating the wonderful biomolecular machinery associated with the promoter

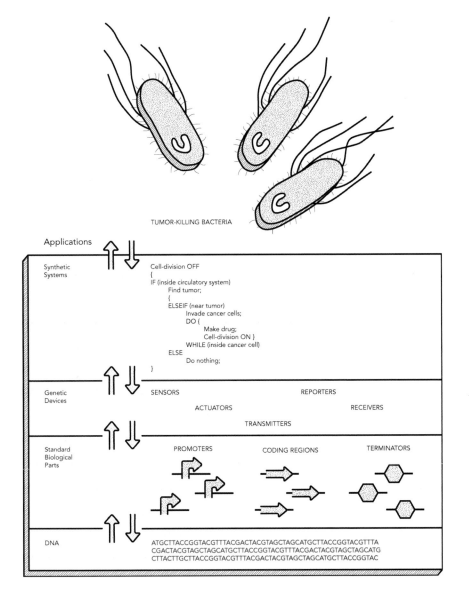

TUMOR-KILLING BACTERIA

Applications

Synthetic
Systems

```
Cell-division OFF
{
IF (inside circulatory system)
    Find tumor;
        {
        ELSEIF (near tumor)
            Invade cancer cells;
            DO {
                    Make drug;
                    Cell-division ON }
                WHILE (inside cancer cell)
        ELSE
            Do nothing;
}
```

Genetic
Devices

SENSORS   REPORTERS

        ACTUATORS   RECEIVERS

            TRANSMITTERS

Standard
Biological
Parts

        PROMOTERS   CODING REGIONS   TERMINATORS

DNA

ATGCTTACCGGTACGTTTACGACTACGTAGCTAGCATGCTTACCGGTACGTTTA
CGACTACGTAGCTAGCATGCTTACCGGTACGTTTACGACTACGTAGCTAGCATG
CTTACTTGCTTACCGGTACGTTTACGACTACGTAGCTAGCATGCTTACCGGTAC

and its initiation of transcription in the image of a faucet, its complexity has been set aside. There is no need to understand the binding of regulators to the DNA, the role of cofactors, the separation of the DNA strand to allow initiation, and so on. This "black-boxing" of intricacy is an everyday notion that originates in the arch pragmatism of engineering. Numerous technologies were introduced to society before the science behind them was understood. We do not possess a full understanding of gravity, the aerodynamics of bird flight, or exactly how a bumblebee is able to fly, but this does not worry the millions of people who travel by air each year. Black-boxing in synthetic biology causes disquiet to many scientists whose careers are invested

Figure 1.6
The abstraction hierarchy in synthetic biology represents the move from DNA sequence, to devices, to applications.

in infinitely deconstructing biological complexity in search of fundamental understanding. Yet, abstraction allows operational understanding of an item at the level appropriate to the needs of people who seek to use the item; a computer user need know nothing of the machine language running on the CPU, the software engineer doesn't have to have expertise in the power supply design, the electrical engineer need not worry about the exact atomic distribution of boron "chips" in the silicon "cookie" of the microprocessor. The BioBrick program is a way to black-box the complexity of DNA coding by abstracting the function of a DNA part, such as the promoter (figure 1.6). The utility of black-boxing and abstraction are predicated on being able to achieve reliable function and composition among the parts. The behavior described in a part's datasheet ought to be correct for all instances when the part is used; otherwise, each instance would need a new datasheet. Unfortunately, the complexity of nature makes realizing this goal hugely challenging. Biological parts often do not behave as expected, and genetic circuits can interact with their cellular chassis in unanticipated ways. There are numerous ongoing efforts to arrive at the robustness of function in our genetic circuits that is desired. If the chassis organism itself could be rendered less complex, then the possibility for interaction would be minimized. Such minimal chassis have been created for simple bacteria, and efforts are under way to repeat this on the larger genome of yeast.[7] An alternative approach is to decouple, or disengage, the metabolic machinery of the chassis from that of the circuit. There are many ways that this could be achieved, such as by using a different DNA type that uses a different coding[8] or transcriptional machinery from a different organism that reads the DNA in an alternative way.[9] There is a great deal of foundational work yet to be done to deliver the engineer's dream of robust predictable function; indeed, it may be that engineers will have to learn to dream differently.

Genetic mutation is the engine of evolution, but how do we achieve reliable performance in a device when there is the likelihood of it spontaneously mutating away from its designated function? This ability is at once a huge challenge for the engineer but also offers a wonderful opportunity. Our new engineering design toolkit leverages past work to develop and apply evolution as an algorithm for selecting optimized genetic designs. Synthetic biologists may thus become intentional schizophrenics; we may implement part of our designs to insulate them from evolution, to protect it from change, while another function within the same design we implement to encourage adaptation toward a defined endpoint. Seeking to capitalize on evolution so as to deliver an optimized design is a technique already used in metabolic engineering. Using a technique called directed evolution, genetic devices are implemented to encourage a diversity of solutions from

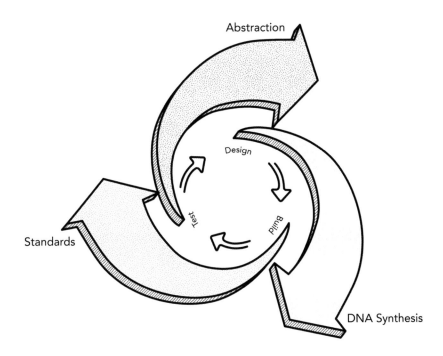

which the best is selected. Insulating devices from change is more difficult. We may seek to put our devices close to regions of the genome, which are known not to have changed much over history. These conserved regions of the genome are usually associated with fundamental life processes, the mutation of which causes cell death. More radically, we may seek to implement a new genetic code that cannot evolve.

A fundamental principle in engineering is the design–build–test cycle as seen reconstrued for synthetic biology in figure 1.7. This approach has underpinned the way in which products have been developed since before the profession of engineer ever existed. A device specification is defined, its design conceived, and a prototype built for testing. The performance information gained from prototype testing is then fed back into the design phase such that an improved prototype can be created. This process is then repeated until a design fulfills the specification desired. The formalization of this approach became the "engineering cycle." The integration of computer-aided technologies in modern making has moved the engineering cycle into a virtual realm. Computational modeling has meant that most iterations of the design–build–test cycle can occur in the computer. The prototype manufactured can be anticipated to fulfill the design criteria, the role of the prototype now being to validate the success of the computational modeling. This has translated into a twenty-first century industry that is capable of conceiving and manufacturing highly complex products with impressively short turnaround times.

Figure 1.7
The design–build–test cycle has been a foundational concept in engineering for many hundreds of years. It remains at the core of modern engineering, although the testing is now often conducted in a computer.

In traditional engineering areas, the development of computer-aided design and manufacture (CAD and CAM) necessarily took place after ways of making had been developed. With synthetic biology, for the first time we are adopting a new family of materials into an age when computational approaches are at the forefront of our thinking. Vast computational power is at our fingertips, but its use is constrained by the paucity of our knowledge of many of the biological processes that we wish to engineer.

What's New? What's the Big Deal?

One perspective on synthetic biology asserts that it is merely about simplifying or deskilling biotechnology; by capturing and sharing the practical art of genetic modification, we render routine the programmed making already carried out by organisms throughout the tree of life. The boundaries of this ambition for synthetic biology lie in the rendition of biotechnology truly into the industrial realm. In defining synthetic biology, we are surely better served by reflecting on the motivations for formalizing the craft of biotechnology: to allow the pooling of knowledge and its gifting to a community of biological engineers. Synthetic biology should not be considered deskilling but rather as the "up-skilling" of biotechnology. The sharing of knowledge and process gives the freedom to focus attention on harnessing the complexity of biology in a manner sympathetic to the incredible capacity for manufacturing exemplified by a pinecone or abalone shell. By capturing a base level of knowledge and craft, synthetic biology aims to deliver a step-increase in the accomplishment of our endeavors. Up-skilling will deliver more than the efficient and rapid realization of our known ambitions, but the next stride, the unanticipated future. When the laser was invented no one had any use for it, but it became central to many consumer electronics and the information superhighway. To frame synthetic biology as subservient to the requirements for industrial biotechnology risks disallowing a universe of unknown applications for this technology.

Where could those unforeseen futures arise? The fascinating capability of biological systems to execute programmed synthesis of very complex molecules is immediately attractive. In addition to making bio-analogs (exact copies of molecules already made in nature), we can make bio-similars, molecules possessing useful attributes that are similar to ones that are made in nature. For example, the terpenoid class of biomolecules contains some 50,000 theoretically possible configurations of which we know some 20,000 are made in nature. From these we extract some 2,000 from plants, animals, and insects, including various things like flavors, scents, antibiotics, dyes, and so forth. There is the strong prospect that within the unmade 30,000 we will find new products.

There is a risk that speculating about and embracing these potential futures may serve to alienate if not done carefully. "We are the victims of bio-tech 'imagineering,'" asserts Eugene Thacker, the "blatant disparity between hyper-optimism and an overall lack of concrete results." All technologies aspire to hold the promise of a better future: faster, cheaper, better products resulting in happier, healthier, enriched humans. Biotechnology was no different, but why is it perceived to have been such a disappointment, to have so grossly undershot its promise? The perception that it wholly failed to deliver "concrete results" is worthy of scrutiny. Today there are in fact many, many examples of products from biotechnology: the protein-digesting enzyme in biological washing powder made by the bacteria *Bacillus licheniformis*, animal rennet substitute for cheese making manufactured by the yeast *Kluyveromyces lactis*, and human insulin synthesized by the *Escherichia coli* bacteria to name but a few. The perception of a failure to deliver is in part engendered by a lack of public awareness of the successes. The products, in the main, are not of themselves genetically modified organisms (GMOs), but have instead been synthesized by GMOs. This permits manufacturers to choose not to draw the consumer's attention to their use of GMOs; for example, from the example mentioned above, the rennet is identified on the cheese packaging as "vegetarian," rather than "synthesized by a GMO."

The discovery of the molecular scissors for DNA, called restriction enzymes, in the early 1970s enabled a form of genetic "cut and paste" in which genes could be removed from one organism and introduced into another. At that time, knowledge of the genome and its functioning was very immature; application of a design–build–test cycle was futile as, more often than not, little could be learned from a failed test. Adoption of a formal engineering approach to genetic modification was not feasible; the ability to cut and paste DNA came along before sufficient depth of knowledge of molecular biology and the genome. The power of restriction enzymes could not be harnessed much beyond the serendipitous. Despite this, much of the biotechnology industry was founded on this ability to create genetic modifications encoding simple declarative statements, like "make drug." In the resulting rush to profit, the early products selected for development were naturally those bearing highest profit margins. In pursuit of such products, those in the vanguard of "genetic engineering" did not invest in creating the infrastructure to underpin what could be considered an engineering industry. With high profit margins, efficiency and lean-running were not principle concerns; being first to market was far more important in ensuring commercial success for a product. This biotechnology version 1.0 persisted across the decades of the 1970s to 1990s, but this commercial model is now in trouble. Blockbuster products, like new drugs with high profit margins, are becoming increasingly rare. For sustained productivity, biotechnology must develop alternative products such as commodity/bulk chemicals, but the industry arguably lacks the engineering foundations necessary to achieve the efficient delivery of a diverse, high-volume, low-cost product portfolio. This situation has created an awkward legacy for twenty-first century biotechnology.

The state of our knowledge has progressed a great deal since the early days of genetic engineering. Dramatic changes in our understanding and capacities have come in three key areas: the reading of DNA (sequencing), the chemical writing of DNA (synthesis), and, more lately, the interpretation of the functions encoded by the DNA. One can think of biotechnology as having opened a door to a version 2.0 during the 1990s to 2000s, as much better control of the flow of molecules through synthesis pathways within cells was learned. Metabolic pathway engineering enabled the refinement of those simple declarative statements of biotechnology version 1.0, taking us from "make drug" to "make a lot of high-value *compound A.*" This was economically viable in the circumstance in which *compound A* still commanded a substantial price per kilogram, such as for the enzymes for detergents described earlier.

Synthetic biology may be understood as having a chance to realize biotechnology version 2.0, with 40 years of rapidly advancing knowledge enabling us to return now to the early vision of biotechnology and believe that we can begin to deliver more fully on that which was imagined at the genesis of genetic engineering. Should this be the bounds of our thinking? The importance of empowering industry to instruct genetic statements such as "make vast quantities of low-value *compound B*" (where *B* could denote biofuels, bioplastics, bio-similars, and so forth) is considerable, both economically and geopolitically. Indeed, the move away from reliance on fossil fuels and materials that this would enable may be of great environmental benefit also. However, we are still in pipes; both the miles of stainless steel tubes in big industrial biochemical plants seen in figure 1.8 and the metaphorical pipes of metabolic pathways. They may be really efficient, very well designed pipes, but they still constrain our ambitions. Synthetic biology should not be about only being better at routing metabolic effort down a pathway in order to make *compound B* more efficiently. Nor must synthetic biology necessarily be restricted to producing products in bioreactor vats in order to achieve sustainable manufacture. Restricting the application of the tools of synthetic biology to the service of industrial biotechnology would be a failure.

To view being "locked in pipes" as being the failure to fulfill a vision is controversial. Delivering the biofuels at the scale needed and at an economically viable price could be a huge success for the field. But there must be a grander vision of synthetic biology in which the designs we implement

become conditional: able to execute complex logical operations. This may be understood as synthetic biology being the third generation of biotechnology, one that takes genetic modification beyond the simple declarative statements of biotechnology version 1.0, the "make drug" decades of the 1970s to 1990s. At that point, when biology becomes an executable programming language, with logical modules and two-way communication with inorganic made objects, then we can say we have left behind the metaphorical pipes of metabolic pathway engineering.

A powerful influence on the direction of synthetic biology is our budding ability to custom-write DNA. Synthesis of DNA is beginning to release us from having to rely on those parts offered to us by nature. We can conceive of

Figure 1.8
The adoption of synthetic biology may well move the chemical industry away from the highly centralized infrastructure of the petrochemical industry. Small-scale, distributed production plants, possessing the flexibility to produce a range of products, might become the norm.

commissioning molecules that have never existed in nature. DNA synthesis has fallen in cost in the past decade to a point at which it is often more convenient to synthesize a whole gene cluster rather than compose it from Bio-Bricked parts. A gene would typically cost a few hundred dollars. If the trend of falling cost continues, there is the potential that in a decade this may be the price of a whole genome. A number of companies offering DNA synthesis services have come into being, or you can buy your own synthesizer (such as seen in figure 1.9). This decoupling of design from fabrication reinforces the abstraction hierarchy and empowers design at a functional level. We are not constrained by selecting the part with the function closest to our needs; we can create the precise part, bespoke. This opens a new age, free from the need to recapitulate biology, the decades of constrained gene jockeying superseded by an era of DNA authorship.

If taken to its logical conclusion, this ability to synthesize DNA prompts a tantalizing question: Do we need to rely on modifying existing chassis organisms or can we conceive of assembling a chassis in its entirety? The potential

Figure 1.9
A twentieth-century DNA synthesizer made by GeneForge Inc., capable of writing short strands of DNA.

to execute a genetic program outside a cell already exists. By combining in a test tube all the machinery needed to sustain one-way information flow through the central dogma (i.e., DNA → RNA → protein), cell-free gene expression is readily achievable. Similarly, we can take biomolecules and construct them into the basic structures of a cell, such as the membrane (its outer skin). Combining these two abilities one can make a protocell— a cell-like object containing gene expression machinery. This is an active area of research in academia with groups seeking to establish the simplest built system that displays life-like behavior; such endeavors are discussed in chapter 15. As attempts to build life "bottom-up" progress in parallel with efforts to simplify organisms "top-down," there is some future where these efforts meet in the middle: Which then would be more natural, the simplified organism or the constructed?

But How Might Biology Change Engineering?

Engineers are beginning to understand how to design and make using biology as a material. Is there learning from this that can be reflected back on how we make using traditional materials? If we can stimulate engineers to dream differently, what lessons from biology could inspire them? Reflecting back to the opening of this discussion, there are certain to be lessons to take from the total recycling achieved by biology. Ecosystems are incredibly adept at reusing their atoms—even those animals at the top of the food chain will ultimately become food for the base of the ecosystem after death. How could products be re-engineered to become entirely recyclable? Can we achieve a future in which we no longer bury a single atom of waste? Probably, if we make it so. Already, the life cycle of some products is being considered such that their component materials are accessible for re-use. In contrast, other products are intended to be transient with no regard given to their fate after being discarded. Such wanton waste within a natural ecosystem would be such a competitive disadvantage that it would soon collapse.

(Raised) Expectations

Is there a risk that we are set to make the same mistakes as those of earlier forms of biotechnology? A great deal has been promised of synthetic biology, but there is still much knowledge to be captured and development needed to generate the underpinning resources necessary to re-engineer biotechnology. The ambition of leaving behind the actual pipes of the bioreactor, leaving a contained environment and launching synthetic biology into the wider environment, into ecosystems, is a further and substantial technical challenge. Our best efforts are currently confined to single-celled organisms; applying synthetic biology to multicellular organisms, such as plants, raises

Figure 1.10
In its 2010 iteration, the iGEM competition saw 130 teams from across the globe compete, with strong representation from Asia and Latin America in addition to North America and Europe.

the bar. Yet it is possible that the technical obstacles will not be the most difficult to negotiate. Use of synthetic biological organisms within natural ecosystems will require extended dialogue around its regulatory and legal implications but importantly will demand extended societal debate on the benefits, risks, and uncertainties of the technology. Scientists and engineers hold a responsibility to consider the impact of the knowledge and technology that they bring to the world. In our imagining, dreaming, and aspiring to create a set of future possibilities, it is clear that as a community, synthetic biologists need to engage in debate with wider stakeholders about the purposes of their work and whether or not these would best be achieved using synthetic biology.

There is an unfortunate, but difficult to avoid, bias throughout this chapter. We consider synthetic biology only for its application in the developed world. A simple analysis of the team nationalities participating in the iGEM competition shows us that synthetic biology is attracting interest from across the globe (figure 1.10). We sincerely advocate that synthetic biology and its deployment are not monopolized by the developed world but equally are not imposed upon others. Synthetic biology ought to be able to offer many applications tailored to local requirements. We refrain from engaging in discussion of these not because of disinterest but rather a desire not to prejudice opinion as to how synthetic biology should be deployed. The enabling power of the up-skilling inherent in synthetic biology can deliver great capacity in research and development to developing countries thanks to the simplification of infrastructure and lowering of cost—biology is the ultimate distributed manufacturing platform.

The emphasis of this introduction to synthetic biology has dwelt on using organisms as a new substrate to implement programmed making. This should not be construed as a statement as to the bounds of synthetic biology's impact; rather, this is to serve the nature of the discussion presented in this book. The tools, which are being developed for the construction of engineered biology, will have a fundamental impact on the way that research is conducted in the life sciences. Bioscience is poised to experience a revolution, the transition from learning by dissection, the deconstruction of complexity in search of simple underlying relationships, to a new constructive age where understanding is generated by synthesizing biology to test hypotheses. This is a transformational shift that promises to initiate a feedback loop; the increasing knowledge produced from synthetic biology will help us to become better at synthetic biology. There is the hope that the twenty-first century will be one in which humankind can learn to work in harmony with biology, to gain new knowledge, create new life, and build a sustainable bioeconomy to support us and all of nature's ecosystems into the next millennium.

2.
Countering the Engineering Mindset: The Conflict of Art and Synthetic Biology

Oron Catts and Ionat Zurr

We overlook only too often the fact that a living being may also be regarded as raw material, as something plastic, something that may be shaped and altered.[1]
—H.G. Wells, "The Limits of Individual Plasticity," 1895

More than 100 years since H.G. Wells wrote his essay "The Limits of Individual Plasticity," it seems that we no longer overlook the fact that life is a raw material for our engineering dreams. Life is increasingly seen as the new frontier for exploitation: from industrial farming through in vitro meat and bio-prospecting to synthetic biology, life is extracted from its natural context and transferred into the realm of the manufactured. This is all part of a larger human project that can be called the *single engineering paradigm*. This is a vision of a future in which the control of matter and life, and life as matter, will be achieved through the application of engineering principles, from nanotechnology to synthetic biology and, even, as some suggest, geo-engineering, cognitive engineering, and neuro-engineering. The central claim of the single engineering paradigm is that it is the application of *real* engineering logic onto life. This is particularly true in the case of synthetic biology. Its rhetoric proclaims that it is not merely rebranding existing forms of manipulating life, but rather that it represents a far-reaching shift in the way life is perceived and used. Such claims are not new, but in recent years we have witnessed a resurgence of the application of engineering logic in the life sciences, coupled with the treatment of life as a raw material for manufactured products. Synthetic biology's proponents claim that it will revolutionize the technology of the future.

Engineers are interested in biology because the living world provides a seemingly rich yet largely unexplored medium for controlling and processing information, materials, and energy. Learning how to harness the power of the living world will be a major engineering undertaking.[2]

Synthetic biology can also be seen as a "catch phrase" of contemporary attempts to apply engineering logic to life, and, as such, it covers a wide range of approaches ranging from the rebranding of genetic engineering to the creation of synthetic life forms (e.g., "Synthia"[3] and/or protocells). Driven mainly by engineers, synthetic biology in parts of Europe and North America is going through an image construction exercise that is interesting to follow. Mechanical engineer turned synthetic biologist Alistair Elfick likes to quote comedian Simon Munnery's gag about molecular biologists: "The engineering equivalent of genetic engineering is to get a bunch of concrete and steel, throw it into a river, and if you can walk across it, call it a bridge." According to Elfick, now that engineers are moving into biology, the construction of the "bridge" will finally be done the right way. However, because of the complexity of biology, synthetic biology is unlikely to ever be as predictable as bridge construction. The bioengineer Frances Arnold notes: "There is no such thing as a standard component, because even a standard component works differently depending on the environment… The expectation that you can type in a sequence and can predict what a circuit will do is far from reality and always will be."[4]

Having control over life and its processes may have always been an ambitious human endeavor. What is changing, however, are attitudes toward life resulting from the accumulation of scientific knowledge and technological capabilities, and the increasing speed and scale of manipulation. A choreographed interplay between hype and actuality is overlaid on a public that is bombarded with information that should excite but is also easily forgotten. It seems that whereas previously, biologists were applying their understanding of engineering to the life sciences, now it is the engineers who force-fit engineering methodologies onto living systems; life is becoming biomatter, waiting to be engineered.

One important aspect of applying this engineering mindset to the manipulation of life is the idea that it would make biomatter easier to engineer, and hence gives to the uninitiated the ability to manipulate and create new life. As a result, life is becoming raw material for artists, designers, hobbyists, and amateurs. Artists and designers are already engaging with biomatter in ways that only a few years ago would have been hard to imagine. However, a worrying trend is that engineers, who are used to a system of certifications and approvals, tend to favor restrictive approaches in regard to access to these new techniques.

The narratives of synthetic biology oscillate between openness and control, global standards and restriction of use, while emphasizing that the unparalleled power to engineer life is waiting for us just around the corner. The term "bioerror" has been thrown around frequently with regard to the combination of ease of use and access to the power of new biological technologies. The idea that the engineering approach will yield full control over life and will do so with unmatched ease seems to be of major concern to safety experts and government agencies that subscribe to the hype of contemporary biological engineering. Controlling access and monitoring developers and users seems to increase at the time when "do it yourself" (DIY) and artistic use of biomatter surges.

Historical Reflections

The application of engineering logic to life has historical precedents. In 1895, H.G. Wells reflected on the body as a malleable entity in his essay "The Limits of Individual Plasticity," writing "[t]he generalization of heredity may be pushed to extreme, to an almost fanatical fatalism."[5] A year later Wells demonstrated some of these ideas and their possible consequences in his novel *The Island of Dr. Moreau.* The plasticity of living processes through real, not fictional, human intervention was demonstrated quite spectacularly only 3 years later in 1899 when Jacob Loeb developed what he called "artificial parthenogenesis ... the artificial production of normal larvae (plutei) from the unfertilized eggs of the sea urchin" (figure 2.1).[6] Loeb had demonstrated the capacity for fertilization (in a sea urchin) without the use of sperm. Following his discovery, he wrote, "it is in the end still possible that I find my dream realized, to see a constructive or engineering biology in place of a biology that is merely analytical."[7]

LOEB TELLS OF ARTIFICIAL LIFE.

Describes Experiments at the Convention of the Western Naturalists.

SAYS SALT IS A FACTOR.

Fish Eggs Developed by the Imparting of a Mild Electric Shock.

MEN OF SCIENCE CONVENE.

Loeb symbolized a change in the field of the biological sciences from descriptive to prescriptive, from the realm of knowledge gathering to the realm of technological application. Loeb adopted in his experimentation and biological research what he described as the "engineering standpoint."[8] His strong belief in control over life and his mechanistic approach to it led him to argue that "instinct" and "will" were "metaphysical concepts ... upon the same plane as the supernatural powers of theologians."[9]

In 1906, Loeb was one of the first to discuss the idea of life as biomatter that can be synthesized. Having a mechanistic view of life, he suggested that biology should shift from mere observation to manipulation/engineering. As a thought experiment, he also suggested making a living system from dead matter as a way to debunk the vitalists' ideas,[10] and he demonstrated abiogenesis,[11] establishing that life and matter are one.

The belief that life is a by-product of matter that can be engineered might lead to interesting interpretations and applications. Ten years after Loeb's work, Alexis Carrel, a surgeon, demonstrated the plasticity of the body through the development of the technique of tissue culture—the growth of

Figure 2.1
Jacques Loeb's discovery of artificial parthenogenesis, as reported in the *Chicago Daily Tribune* in December 1900.

living tissue/cells in vitro—in an artificial environment. Carrel was a well-known and respected scientist who had developed new medical techniques in suturing arteries and transplantation as well as tissue culture and won the Nobel Prize in Physiology or Medicine in 1912. He was also a complex and controversial figure—a person who pushed the implications of his discoveries to some extreme and morally questionable places, far from the strictly biomedical or even scientific realms into ontological and sociopolitical issues.

In the 1930s, Carrel, *the surgeon,* joined forces with *the mechanic*—the famous aviator Charles Lindbergh—to devise the Organ Perfusion Pump. This mechanical pump circulated nutrient fluid to feed large organs kept alive outside of their host body. Carrel's affiliation with Lindbergh, the great American hero, extended to a shared ideology of eugenics, which Carrel outlined in his 1938 publication *Man, the Unknown*: "Those who have murdered, robbed . . . kidnapped children, despoiled the poor of their savings, misled the public in important matters, should be humanely and economically disposed of in small euthanasic institutions supplied with proper gases. A similar treatment could be advantageously applied to the insane, guilty of criminal acts."[12]

It can be argued that the application of mechanical/engineering logic to the living body followed the same line of thinking that led Carrel to treat human societies as objects to be engineered. These engineered objects can be fixed by removing faulty parts. "Eugenics," Carrel wrote in the last chapter of *Man, the Unknown*, "is indispensable for the perpetuation of the strong. A great race must propagate its best elements."[13] The book was a worldwide bestseller translated into 19 languages.

Following Loeb's dream, in 1952 Stanley Miller and Harold Urey managed to create amino acids in a test tube, simulating the conditions of the early Earth (figure 2.2). In 1955, Heinz Fraenkel-Conrat and Robley C. Williams showed that a functional virus could be created out of purified RNA and a protein coat. At the time, this work was described as "life created in the test tube." In 1965, Prof. Sol Spiegelman and his team at the University of Illinois "succeeded in putting together the non-living nucleic acid message which produced a virus," creating what was nicknamed "Spiegelman's monster," and in 1967, "Arthur Kornberg at Stanford University and colleagues announced that they had copied the DNA of the Phi X174 virus, producing an entity with the same infectivity as the wild virus."[14]

In 2010, Craig Venter declared that he created the first life form whose "parent is a computer."[15] Even though what Venter achieved is indeed a great technological feat—being able to replace the genome of a simple bacterium with a synthesized version—the context (i.e., the cell and its contents) is still of biological origin. The role of the computer in synthesizing the genome was that of a glorified copier—replicating a genome without fully understanding

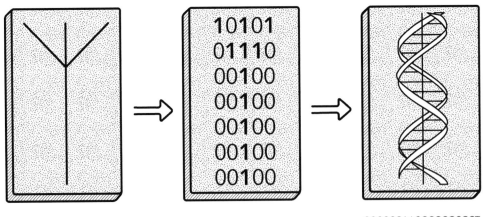

Microvenus icon

CCCCCCAACGCGCGCGCT

its meaning. The only novel part that was introduced to the genome was a "watermark"—a hidden message that contained the names of the researchers, a URL, and quotes relating to creation of life—something that the artist Joe Davis did back in the mid-1980s,[16] showing that artists can precede scientists in conceptual breakthroughs (figure 2.3).

Cultural Reflections

The concept of synthetic life (life engineered from raw materials) presents one of the most challenging, deeply philosophical, ethical and cultural areas for artistic scrutiny. Promising to fulfill Loeb's dream of debunking the vitalists' ideas, and used by proponents of "intelligent design," synthetic life is ready to assault cultural sensitivities about life. However, synthetic life is only one of the areas now labeled synthetic biology. The area where most of the "real" engineering activity is finally applied to genetic engineering is the standardization of biological parts. One of the most famous events associated with this ambition is iGEM—the International Genetically Engineered Machine competition. This is a competition in synthetic biology for undergraduate students, where the teams use and develop standard biological parts, known as "BioBricks." iGEM has grown quite dramatically—from five teams in its first year (2004) to more than 160 in 2011. Relevant here is that one of the iGEM judges in 2009 was an artist: Richard Pell. Furthermore, in 2009 a team from the Srishti School of Art, Design, and Technology (Bangalore, India), won a bronze medal and the prize for best scientific poster (see chapter 6).[17] This team was made up of design and art students; none were scientists or engineers. Their project involved the (poetic) creation of bacteria that produce the smell of earth after the rain. In 2009, the winners of the grand prize, the University of Cambridge team,[18] worked with designers, including Alexandra Daisy Ginsberg, who acted as project advisors (see chapter 6). By a strange coincidence, 2009 also saw for the first time the very public presence of the FBI at the competition. FBI agents engaged with the teams and urged them to work closely with government agencies and safety officials.[19] iGEM in 2010 saw more teams working with designers as advisors as well as two artist teams, including a new Srishti team with a project that demonstrated how the engineering of one organism influences the phenotypic behavior of another.

Synthetic biology seals the entry of the engineering mindset into the life science laboratories, at the very same time that the artistic mindset has invaded these spaces. The scientific laboratory where artists (and engineers) are placed cannot be viewed in isolation but is positioned within the larger system of our times, one that calls for utility, efficiency, and profit. It is also a time of ecological crisis and its gloomy future predictions.

Figure 2.2
Stanley Miller demonstrating the famous Miller–Urey experiment, which simulated the chemical conditions of the primordial Earth's atmosphere.

Figure 2.3
Depiction of the first nonbiological message encoded in DNA. In 1988, Joe Davis, an artist collaborating with molecular biologist Dana Boyd in Jon Beckwith's lab at Harvard Medical School, designed and synthesized an 18 base-pair message encoding the image of the ancient Germanic rune representing life and the female Earth. The *Microvenus* message was then pasted into a vector and transformed into *Escherichia coli*.

The Engineering Mindset

In an article that received relatively wide coverage, sociologists Diego Gambetta and Steffen Hertog demonstrated that "among violent Islamists, engineers with a degree, individuals with an engineering education are three to four times more frequent than we would expect given the share of engineers among university students in Islamic countries."[20] After eliminating other plausible possibilities such as network links and/or technical skills, they conclude that an "engineering mindset" as well as current social conditions in Islamic countries as the most plausible explanation of their findings.[21]

First, we should highlight, as emphasized by the authors of the paper cited above, that the engineering mindset is not unique to Muslim extremists. On the contrary, it is a mindset that characterizes extremist thinking in general, engulfing all religions as well as secularism. Second, this mindset can be described generally as one of the fundamental views of the world: the either/or, right/wrong attitude. Furthermore, "[t]he concept includes an assumption, which has been raised in psychological research, that engineering as a field of study and a profession tends to attract people who seek certainty, and their approach to the world is largely mechanistic. So they are characterized by a greater intolerance of uncertainty—a quality that is evident among extremists, both religious and secular."[22]

A Case Study

Although not operating at the molecular level, the following can illustrate some of the problems assisted with the transformation of life into a material to be engineered. Living cells and tissues are removed from the original context of the body to be used as raw material, such as in the development of in vitro meat. Human relationships with meat tend to reflect broader cultural attitudes toward complex living systems. Where eating animals once had cultural and symbolic significance, such as acquiring the attributes of the animal consumed, now "factory farming considers nature as an obstacle to be overcome."[23] Factory farming represents the industrial-scale application of engineering logic onto animals in order to maximize production. This approach is starting to be seen as archaic and problematic, less on ethical grounds, but more for its perceived inefficiency and environmental cost: "alternatives can be produced under controlled conditions impossible to maintain in traditional animal farms, they can be safer, more nutritious, less polluting, and more humane than conventional meat."[24]

The idea of growing meat without the animal has been around for some time in science fiction (described in *The Space Merchants*, 1952). Even Winston Churchill wrote in 1932, "We shall escape the absurdity of growing a whole chicken in order to eat the breast or wing, by growing these parts

separately under suitable medium."[25] It was only in the late 1990s that this idea started to be considered seriously and investigated technologically. With a new approach to the body manifested through regenerative medicine, in vitro meat simply represents a nonmedical application of tissue engineering principles (figure 2.4).

As part of The Tissue Culture & Art Project (TC&A), we developed a series of pseudo-utilitarian works under the banner of *Technologically Mediated*

Figure 2.4
Victimless Leather: A Prototype of a Stitch-less Jacket Grown in a Technoscientific 'Body' (2004), made of biodegradable polymer and connective and bone cells by the Tissue Culture & Art Project.

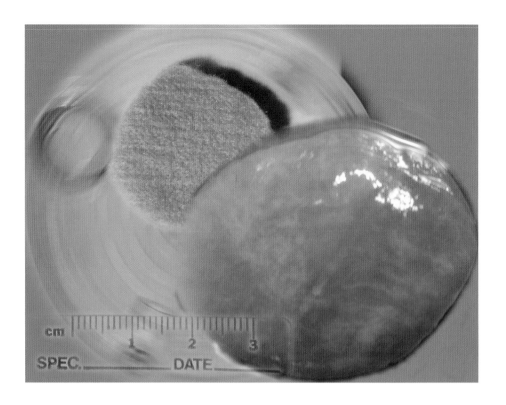

Figure 2.5
Tissue Engineered Steak No. 1
(2000). This Tissue Culture & Art
Project study for *Disembodied
Cuisine*, made of prenatal sheep
skeletal muscle and degradable
PGA polymer scaffold, was
part of Oron Catts and Ionat
Zurr's Research Fellowship in
Tissue Engineering and Organ
Fabrication, MGH, Harvard
Medical School.

Victimless Utopia that included the first ever growth and consumption of in vitro meat (figure 2.5). Starting in the year 2000, TC&A has grown muscle tissue as meat. In 2003, TC&A presented a performance/installation titled *Disembodied Cuisine*, in which frog skeletal muscle cells were grown over biopolymer, and healthy frogs lived alongside as part of the installation. In the last day of the show, two steaks were cooked and eaten in a Nouvelle Cuisine–style dinner; four frogs, rescued from the frog farm, were released to a pond at the local botanical gardens.

Even though our artistic project was set up to highlight the irony of transforming meat into the ultimate engineered matter, in vitro meat is now considered seriously as a potential alternative for the production of animal proteins for human consumption. One complication arising from the victimless meat endeavor as a manifestation of the technoscientific project is that it creates an illusion of a victimless existence. In vitro meat needs a serum created using animals' blood plasma. Although research to find alternatives is under way, there are currently no solutions in the near term, and animals are killed for it (mainly calves for fetal bovine serum). Also, all the "costs" concerned with the running of a laboratory (i.e., fossil fuels burned, greenhouse gases produced, water and trees consumed, miles traveled and the waste created) means that the ecological footprint of in vitro meat is significant.

In addition, there is a shift from "nature red in tooth and claw" to a mediated nature. The victims are pushed farther away; they still exist, but are much more implicit.

Animal cells cannot manufacture nutrients from nothing; in vitro meat is merely an engineering exercise in translating/synthesizing nutrients from other sources. In other words, parts of the living are fragmented and taken away from the context of the host body (and this act of fragmentation is a violent act) and are introduced to a technological mediation that further "abstracts" their liveliness. By creating a new class of semi-being, dependent on us for survival, we are also creating a new class for exploitation, as it further abstracts life and blurs the boundaries between the living and the nonliving.

We are now experimenting with muscle tissue as semi-living labor, growing skeletal muscle cells to become actuators. This project further problematizes the distinction between the living (whether human or other animal) and the machine (whether living, i.e., woman, slave, or animal, or nonliving), highlighting the absurdities inherent in notions of ideal technological solutions and efficiencies.

Conclusion

If in recent years the engineering mindset is being introduced into the life science laboratories, so too is the artistic mindset, as in the case of SymbioticA—the Centre of Excellence in Biological Arts in the School of Anatomy and Human Biology at The University of Western Australia.[26]

Criticizing synthetic biology and refusing to take part in its developments may seem to suggest a call to halt the research and its implementation and return to an unattainable romantic past in which humans and nature lived in harmony. This is an unhelpful reactionary perspective. Entering the places where synthetic biology is being developed and using it to create artworks that have the potential (even against the artist's will) to become cogs in the future propaganda machine is also problematic. Yet we believe that artistic expressions that are more subtle and complex will be able to offer glimpses of the possible and the contestable; works that are neither utopic or dystopic but rather ambiguous and messy, acting to counter the engineering hubris of control. We also hope that artists will be able to present/empower those outside the hegemony and give them the voice they deserve.

A future dominated by the single engineering paradigm might be upon us; biomatter is increasingly used as raw material. If this is the case, the engineering approach should not be allowed to monopolize life. One way to emphasize and attract attention to alternative frames of thought is to open up the very same tools and spaces that serve this future to other disciplines, including artistic.

3.
Design as the Machines Come to Life

Alexandra Daisy Ginsberg

Synthetic biology's champions describe a technology that will not only change the way we live and the world around us, but one that might even "save humanity," a "green" weapon against looming threats of energy shortage, disease, hungry populations and climate change. Theirs is a dream of a sustainable future, powered by a human-designed biology, controlled by the logic of engineering. While this promised technology may still be a science, its enthusiasts hope that it could help repair our troubled industrialized landscape: Humanity's needs and consumer desires could be neatly balanced with our planet's limited resources, as our machines come to life.

Frontispiece
Greening Biotechnology.

The big promises attached to new technologies often fail to materialize in quite the way we imagine, but they can still affect how we see the future. This envisaged easy-to-engineer biotechnology is no different. Synthetic biology is attracting not only scientists, engineers, governments, social scientists, and investors, but also a growing influx of artists and designers, keen to understand better synthetic biology's potential implications for the world we live in. While it may not change the world, synthetic biology could give us ways to change the way we think about it.

Experiments, Artifacts, and the Design Perspective

As one of these artists and designers working with synthetic biologists, I have learned that the same words can convey very different, even opposite meanings. In art and design, I use the "experiment" as an open-ended process to open up and reveal potential ideas; in science, the "experiment" is a tool to generate data to test a hypothesis. Repeating an experiment and achieving the same results is key to the scientific method, whereas the experimental process in art often seeks out the exceptional or unique. Artifacts may emerge from experiments. In science, the "artifact" is an outlying bit of data—an erroneous, often human-induced thing that can be ignored, like the distortion caused by the curvature of a lens. Conversely, for the artist or designer, the artifact is the focus of our attention: We are actively *making things*.

But it is another shared word that has made synthetic biology so intriguing to me: "design." In synthetic biology, civil and mechanical engineers, biologists, computer scientists, chemists, and mathematicians all talk about "design." Some even describe themselves as designers. Their molecular blueprints may be executed at the genetic scale, constructed by invisible living machines, but just as biology is bigger than the sum of its parts, so too is their vision.

This seems to be the true novelty of synthetic biology: a field of technoscience proposing, as design does, to make things, rather than focus on understanding existing ones. The engineering ideals of standardization, abstraction, and decoupling are intended to make it easier, quicker, and cheaper to manipulate biology, with less scientific knowledge. If achievable, the dream of deskilled biological design hints at shifting boundaries to come, as the design of applications, rather than DNA, becomes the bioengineer's focus. With designed organisms that will supposedly target cancers, produce novel foods, act as data storage devices, self-repair buildings, or even detoxify the ground beneath us, I see synthetic biology being presented as a design discipline of the future, modeled on—and validated by—design, the existing business of producing stuff.

As I became entangled in synthetic biology, I had to learn as I went along. Even if the medium was exotically biological, there seemed to be affinity in "design"; this word that I assumed had common meaning. But like "experiment" or "artifact," "design" takes on different meaning in synthetic biology than the design I know. We each use the term to meet our discipline's very different ends. Although synthetic biology is described as an engineering discipline, it includes scientists, so language and attitudes to design vary even within the field. Engineering is focused on necessity: solving a defined problem, like designing the structure of a bridge. Science, in contrast, studies life—or the world we live in—as it is, from complex proteins interacting to the forces causing those interactions. But design, as I know it, is different. Design, whether of a building or a chair, is about possibility, experimenting with life as it could be.[1] Architects design buildings that will frame life that has not yet happened. Design projects into our future, creating new possibilities out of existing matter.

Working at the blurred edge of design and art, I investigate what design is and does, curious as to what it can do beyond just translate technologies into things for us to consume. When I first encountered synthetic biology, the language emphasized simplicity: Lego-like BioBricks, aspirations of standardized plug-and-play systems and drag-and-drop design interfaces, all accessible to the non-biologist. Researching the role of design in a biotechnology revolution, I was speculating on what designers might be doing in this century, who they would be, and how they might be trained. With a "Registry of Standard Biological Parts" already cataloged in a freezer humming away at MIT, supplying thousands of undergraduates apparently churning out new parts and applications for the International Genetically Engineered Machine (iGEM) competition, this novel biotechnology sounded tantalizingly close to fruition. I was curious from a design perspective: If biology could indeed be transformed into both machine and material, what were the disruptive possibilities that could affect everything from materials to manufacturing systems to aesthetics? What were the unknowable implications of an unstable—and hence possibly destabilizing—technology on our lives?

Digging deeper, it seems that synthetic biology's headline rhetoric addresses "humanity's needs," rather than our needs as individual, diverse and complex humans, within a diverse and complex ecosystem. Synthetic biology is projected as disruptive technology that also promises to disrupt nothing (figure 3.1). Pumping out limitless "green" jet fuel to feed planes or designing bacteria that secrete the same non-biodegradable plastics that already trouble us addresses neither the failures of our existing infrastructure nor entrenched attitudes to the ecosystem and our place in it. This simply substitutes existing mechanical machinery with biotechnological processes: manufacturing the same liquid fuel in vats, rather than extracting it from the earth. Such heady pursuit under way discourages reflection and cultural analysis of the unique issues and novel design opportunities a living technology presents.

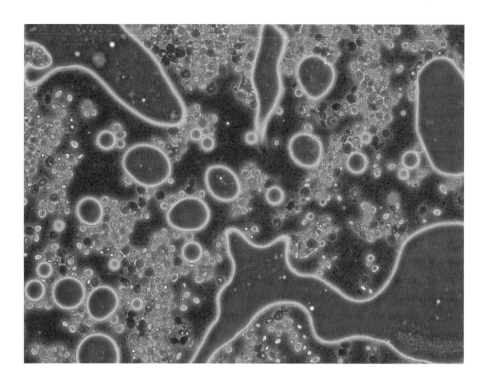

This chapter uses design as a critical lens to understand these aspects of synthetic biology better. To engage this design perspective, I first investigate how design got to where it is and what it means in its more "traditional" practice ("From Nature to Design"). Then in the section "The Redesign of Life," I consider how nature—and especially biology—is already more entangled within the designed products of culture than we tend to think. We then can examine how synthetic biology approaches the question of design, straddling the divide between nature and culture as it shapes biology ("New Designs on Biology"). The simplifications of the Lego analogy break down amid the complexity of cellular networks. If synthetic biology is truly a disruptive technology, then clinging to existing analogies of mechanical and electronic technology may hinder sustainable, ethical, and imaginative design, if that is what we seek.

Although an outsider to science and engineering, I wonder if biology can keep up with the dreams of synthetic biology. My initial reactions to the technology have endured. The seductive appeal of clever science is contrasted with an irrational unease that this technology somehow interferes with the "natural" order of things, amid concern over the potential long-lasting social, political, and physical effects of designing objects, systems, and machines with the very same stuff that we are made of. These issues are raised in the penultimate section "Machines for Living and Living Machines." How do we begin to unpack the social, political, scientific, and cultural complexity of this design vision? It is these areas of uncertainty that make synthetic biology's implications more interesting to explore than the much-peddled narratives of world salvation or biological apocalypse and, as discussed in the previous chapter, make it an interesting partner to art, with its ability to reveal and revel in the flawed.

Synthetic biology may aim to design, but as we will see, it draws on engineering design, the realm of problem solving, not the design practiced in disciplines as diverse as fashion, architecture, or communication. Oron Catts and Ionat Zurr critiqued the engineering mindset in chapter 2, but the contemporary design mindset needs scrutiny as well. The system of industrialized production that incorporates contemporary design means that many designers also find themselves engaged in problem solving, perpetuating unsustainable processes without really being able to address them. Emerging and unorthodox design attitudes, such as "problem finding" and "problem making," which uncouple design from commercial production, are challenging this position, with potential to offer paradigm shifts in the way that we design and engineer things. These may provide the revolutionary moments we seek. Their possible relevance for synthetic biologists' design ambitions will be explored in the final section of this chapter, "New Models for Biological Design."

Synthetic biology does not just present ethical problems to be resolved or technical solutions to existing problems, but fundamental dilemmas with no one answer. Synthetic biologists—and designers—need to design differently to seek a common good, one that does not imply saving a broad-brushed "humanity." Investigating the roles that design could play within synthetic biology, in the process can we also challenge attitudes to how and what we design? Tackling synthetic biology from a design perspective, perhaps we can do what design as a discipline does best: establish possibilities and, through artifacts, experiment with "life as it could be."

From Nature to Design

The verb "to design" has come to define, in some ways, what it is to be human. Language helps us to communicate ideas through words; design translates ideas into things. Design in its most basic sense helps separate what we make from what already exists: natural menaces, living and nonliving, from which we must protect ourselves to survive. But how do we understand design today, beyond a process of translation between concepts and technologies into tangible things or experiences?

Simply defined, design is the act of planning and then making something. This creative process encompasses the way that things or ideas are conceived, the way they are made, the way they look, and the way they function.

Originating in the Latin word *designare*, "to designate," design implies a collaborative, hierarchical, or linear process of production.[2] Designing is not the same as creating; things are not made from scratch.[3] Designed things are a synthesis of ideas and values.

In the past two centuries, the engineering vision of the Industrial and Information Revolutions has transformed the world we live in. Crucial to that process has been design, as it matured into a collection of distinct disciplines that today shapes much of our experience of the world. It was from the grime and smoke of the nineteenth century British Industrial Revolution that designers first emerged, separating their role from that of the stonemasons building or craftsmen weaving. The wholesale change in our lives and environment as Western cities grew, dominating tracts of wild landscape, was synonymous with the emergence of mass consumption. Suddenly, industry's great machines and its workers churned out more stuff for us to consume than our forebears could ever have imagined. Designers helped differentiate this stuff from that produced by competitors, while assembly lines and the division of labor kept it uniform. Humans became consumers to give purpose to the machines' function, and the mass of stuff marked "progress" in our own lives.

Design was integral to the transition from living technologies (horses as transport, clothing from plant fibers) to a world of the nonliving, like sports cars and nylon stockings—the products of mass production, of desirable uniformity and uniform desires. Design gives form to the functions dreamt up by scientists and salesmen. Combustion engines fuelled cars; industrial springs inspired new archetypes like the adjustable desk-lamp; transistors powered personal computers. New technologies inspired the design of new products; new products demanded new technologies to enable their design. Design is so integral to the mechanisms of our consumer economy that since the 1920s, strategies of obsolescence—products designed to fail—have only helped to perpetuate our desire to consume more.[4]

Design, once separate from function—the realm of the engineer—was concerned with just the look and form of things. Now design permeates every stage of the process of making things. Design critic Deyan Sudjic describes the "language of design as the genetic code of our society."[5] By this logic, our ability to make plans and put them into action, a process that underpins societies' cultures, economies, religions, fashions, politics, and pretty much all other endeavors, is design. We are all designers now: designing information architectures, cities, communications, shoes, political revolutions, military campaigns, experiments, and even living organisms. Design itself is uncoupling from the physical object as design thinking and design management gain status as new paradigms of business innovation strategy. By these measures, in English at least, we lack the words to describe such a breadth of activity. Design, as philosopher and anthropologist of science Bruno Latour argues, has come to stand in for anything that is "planned, calculated, arrayed, arranged, packed, packaged, defined, projected, tinkered, written down in code, disposed of and so on."[6]

Future designers of functional living machines—plants, animals micro-organisms—will be descendants of plant and animal breeders, genetic engineers, mechanical engineers, and scientists, but they will also claim heritage from design. Making biology easier to engineer is a driving aim of synthetic biology. To meet this goal of deskilling the biological design process, synthetic biologists already liberally borrow from design. In architecture, design, and engineering, computer-aided design (CAD) software has become ubiquitous; "bio CAD" is now an emerging software market, intended to facilitate drag-and-drop design of DNA. If this is achievable—which is much debated—how much would a biological designer need to know about biology to design it well? Even if biological expertise remains essential, will future biological designers have more in common with today's designers, scientists, or engineers? This depends in part on what synthetic biologists understand by design itself.

Synthetic biology and design today may both be concerned with function, but while synthetic biologists design, they are not "designers" in the same sense. Designers are focused on our interactions with objects and their function, generally operating at a bigger, user-centric scale. The designer is better equipped as a generalist, in contrast to the scientist (and in synthetic biology, the engineer, too), who is a specialist, an expert in the detail of how things work, not whom they work for. The designer's elasticity enables translation across scales and industries to get stuff made. A fashion designer needs to understand cloth, pattern cutting, supply chains, manufacturing, marketing, and have the cultural knowledge that informs the creative design process. Architects learn history, but they also learn basic structural engineering to enable them to collaborate effectively with structural engineers: This is a form of "deskilling," enabling adaptability.

Just as design no longer has a single meaning, there is no universal design process. For bioengineers, design is the means toward a practical solution. It is problem solving defined by efficiency and necessity, a balance between parameters such as cost and function, and interpreted through design pipelines, cycles, and endpoints. Design works differently. It may sometimes be communicated as a distilled, linear path, like "Discover, Define, Develop, and Deliver," but the reality is messier.[7] Designers respond to a brief through research and also think through making: sketching and prototyping, producing unexpected ideas along the way. Live review, integral to architecture education and other design disciplines, comes in the form of critique (the "crit" or "charrette") where work in progress is debated with a panel of critics. While a final design may appear polished, good design itself remains an open-ended process; the unexpected is encouraged. For the engineer or scientist, design's lack of verifiable solutions and its enthusiasm for subjective value judgments, compared to the objectivity projected by the scientific method, may even be alarming. This is not to suggest that science and engineering are not inherently creative processes. As Part Three of this book will show, the Synthetic Aesthetics residents recognized in each other's work the uncertain, winding processes of discovery. Rather, will future synthetic biologists need to be versed in both bioengineering and also human-scale design? Architects may once have designed the structure of their designs but now collaborate with engineers; perhaps product designers will work with synthetic biologists to help design the function of biotech products or even organisms themselves, like the IDEO "living" packaging suggested in chapter 9.

As we seek the limits of what we should design using synthetic biology, what we design is as important as how it is designed. Design innovator John Thackara notes, "We tend to think of products as lumps of dead matter: inert, passive, dumb. But products are becoming lively, active, and intelligent. Objects that are sensitive to their environment, act with some intelligence, and talk to each other are changing the basic phenomenology of products— the way they exist in the world."[8] Thackara is referring to electronic products, but enabled by synthetic biology, lively products may become living ones. This may mark the paradigm shift between synthetic biology's designs and what has come before. As we redefine our interactions with living things, we will need to develop a design discourse around the cultural function and the design itself of biological products.

Such appraisal is almost absent in mainstream design today. Design ethics and criticism are underdeveloped, despite the saturation of design in our everyday lives. Design operates in a complex tangle of other industries and expertise, part of a wider cultural system of divided production that we have come to accept as the norm. As professionals, designers are generally not the owners of the ideas they generate, nor accountable for the functions to which they give form.[9] They work to a brief given by the client, whether a dress, museum, or banking system. The designer delivers the best design within the constraints, loyal to the brief. As a service, the work is less about the designer's personal values, but rather about identifying with the values imbued in the project. Critique of the logic of the system at large is in normal circumstances beyond their remit, for designers as well as scientists and engineers.

Nevertheless, common to many designers, scientists, and engineers is a motivating optimism, a belief that their work in its small way can contribute to making the world a better place. It need not be a cure for cancer;

a well-designed milk carton can be life improving, too. With its instinctive ethical imperative to "do good," how has design been subdued into a service industry sometimes detached from its more humane ideals?

One reason may be because as designers, manufacturers, governments, and consumers, we view the products of culture as somehow independent from the natural world. We see the human touch as transformational, describing the things we craft from nature's materials library as "man-made," "artificial" or "synthetic," that is, synthesized from different things, but even this perceived separation of ourselves from nature is itself a cultural construct. The ecological crisis of climate change shows that human activity and design are never independent from nature.

Today's products before and beyond their functional lives—from the chemistry of their component materials to their life beyond disposal—are not considered the consumer's, nor the designer's, responsibility. In faraway places, others dig holes to extract raw materials, which somewhere else are irretrievably converted into consumables. Once a product's useful life is over, we relinquish it, its toxic components sent back to the ecosystem out of sight, incinerated, buried, or prolonged through recycling. Products are still conceived of in terms of life spans, not life cycles, disregarding knowledge of limited resources.[10]

Perpetuating this mindset that begins at purchase and ends with disposal, consumers are increasingly enticed to enjoy disposable pleasures. For example, the textile industry is one of the most polluting on Earth; it spews out more and more on-trend clothing intended to last a season, while these fashion seasons themselves are ever shortening. Objects are becoming more difficult to deconstruct and repair, or just uneconomical to do so, reinforcing a replacement culture. Amazingly, our definition of good design still includes all these characteristics; parameters of aesthetics, cost, profit, utility, and desire dominate. In the future, good design may mean taking into account long-term thinking, rather than pursuing short-term need and problem solving. Synthetic biology appears to be fitting into the existing systems of design, but we could challenge this. The question whether the ethical burden of the designed object lies with the consumer, designer, manufacturer, or shareholder remains as neglected yet relevant for design as it will be for synthetic biology and the design of living machines.

The relegation of responsibility over what we make is perhaps a curious remnant of attitudes contemporary with the beginnings of consumer society in the late seventeenth century, even before industrialization. Then, blank patches on the world map still indicated territory rich for seemingly limitless exploration and exploitation. With centuries of change behind us, our planet now extensively cataloged, we stubbornly cling to antiquated visions of plenty and a renewable world rich with infinite resources. The birth of modern science, and with it the concept of "progress" that is imbued in modernity, has only helped to enforce a cultural emancipation from nature, placing technology in opposition to the creations to the natural world. Our current ecological situation may well be symptomatic of this phenomenon.[11] Change must come from all parts of the system to be effective, not just from design or technology.

The Redesign of Life

Addressing this perceived need for novel approaches, synthetic biologists are promising a century defined by an engineered biology with the transformational potential of the Industrial and Information Revolutions. Biology—and life with it—will be remastered for the design and construction of useful things. As "design" has come to describe most human activities, synthetic biologists have easily adopted it. But something as significant as designing life should not merge unquestioned into design's sprawl. Synthetic biology will shrink the gap between what we make and what we are, merging our neat categories of nature and artifice.

Instrumentalizing life in itself is not new territory. We may perceive nature and culture as separate entities, but natural and artificial materials have long intertwined in designed objects in a far more complicated narrative. We already design with nature, and specifically biology, in many ways. Before we consider synthetic biology's design ambitions for biological things, we should address the ways in which things are already made from biology.

Harvested plants and slaughtered animals are manufactured into designed objects. Natural cellulose fibers provide our material staples, from cotton and flax for clothing, to wood for construction and paper. Some natural, biodegradable materials are now luxuries compared to their synthetic stand-ins: from furs and leathers like cowhide to stingray shagreen—graspable even when slippery with sweat on the hilt of a sword—to rubber and cork from trees, protein fibers secreted by silkworms, cashmere wool from goats, horse hair, and bone (figure 3.2). All these biological materials supplement a nonliving "natural" palette including stone and clay, and all are shaped by the artificial: technologies from spinning, tanning, firing, to bleaching pervade what we describe as "natural" materials.

In 1856, as mass-produced steel rolled out of the steel mills, 18-year-old William Henry Perkin was working away at his lab bench, tasked with synthesizing the antimalarial chemical quinine. Instead, he accidentally invented the first synthetic dye, mauveine. This was the beginning of the era of synthetic chemistry and with it the arrival of a new library of materials for modern society. Such milestones are not only driven by need or accident, but also by desires. In the 1930s, the holy grail of synthetic textiles—artificial silk—was finally obtained, prompted by drivers of political necessity and the fashion for silk stockings

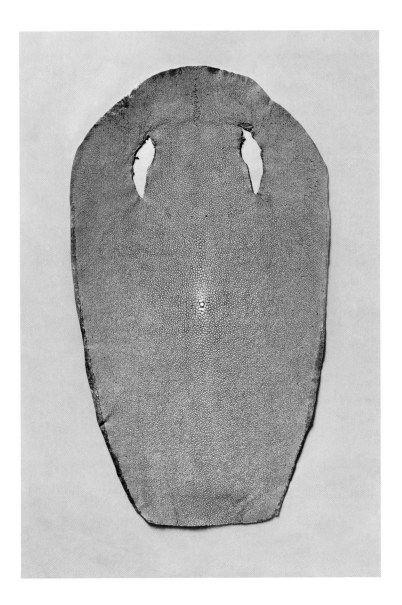

(among other silk products). Importation of Japanese silk was causing diplomatic headaches with the outbreak of World War II; DuPont's Nylon was at last the solution to the problem of unpatriotic hosiery. Nylon joined many other synthetic products, mostly petroleum-based polymers poured from test tubes that were seen as bettering nature. By the 1950s, synthetic chemistry came to define progress as glossy plastics and space-age living, enhancing our cultural appreciation of mass-produced uniformity. But the oil fueling these laboratory alternatives to natural materials is itself "natural," made of dead organisms accumulated over millions of years. Similarly, chemical components of many of the drugs we use were found by "bio prospectors," plant hunters seeking

Figure 3.2
A rayfish skin, ready to be transformed into shagreen-covered books, wallets and even tabletops.

out natural compounds in the wild. These molecules are then copied chemically. Despite the revolutionary impact of synthetic chemistry and its crude oil alchemy, biology remains the source for our materials library.

If the separation between artificial materials and their natural counterparts is complicated, so too is the opaque use of biological materials within "artificial" products. To investigate this, designer Christien Meindertsma tracked all the products made from one carcass, Pig 05049, an animal dismembered and converted into a wealth of materials from pork chops and bacon to ammunition, train brakes, automobile paint, soap, heart valves, bone china, cigarettes and hair conditioner, thereby revealing how little so many of us know about the living origins of the things we consume every day.[12]

When it comes to food, it seems easier to differentiate between processed and natural. Scanning the aisles in the supermarket, the red and juicy tomatoes and the crunchy, sunshine-yellow corn embody nature, plucked from sunny vines or the rolling cornfields printed on the tins. Chewing away,

Figure 3.3
Prehistoric (from left) and contemporary maize cultivars reveal the human design influence of thousands of years of farming.

Transgenic American Chestnut Tree
American Chestnut Tree

ₓ1776
E. coli K-12

Sterile Male Screwworm
Cochliomyia hominivorax

Biosteel™ Goat
Angora Goat

most consumers are unaware that these plants are objects of design, unless they carry a label alerting them to genetic modification (GM), mandatory in the European Union since 1997.[13] Tampering with this classification between natural and "synthetic" food causes anxiety. For some, such an intervention violates the integrity of a "virgin" nature that must be preserved intact. Genetically modified organisms threaten to jump the divide between what should and shouldn't be altered by humans (this of course is not the sole objection to GM).

But even non-GM corn and tomatoes are designed. Humans have been cultivating and domesticating living species for at least 10,000 years for economic as well as aesthetic reasons, whether for pleasure, culture, or intoxication.[14] There are 50,000 known species of edible plants on Earth, but 75% of the world's food intake now comes from just 12 plant species and five animal species. And of these, just three—maize, wheat, and rice—provide more than half the world's calories from plants. According to the United Nations Food

Figure 3.4
PostNatural Organisms of the Month, from the Center for PostNatural History, an artistic research project that investigates the relationship between culture, nature, and biotechnology and documents biological specimens that have been subject to human design.

and Agriculture Organization, 75% of crop genetic diversity has been lost worldwide since the beginning of the twentieth century, as agriculture has become increasingly industrialized. Farmers have abandoned local variants for "genetically uniform, high-yielding varieties."[15] A contemporary corn cultivar compared to its prehistoric ancestors reveals this massive redesign in just its size, shown in figure 3.3.

Nutrition, utility, and flavor are altered too as we design with artificial selection, optimizing living things to meet our intentions and desires. Turning cabbages into broccolis and cauliflowers, and cows into high-yielding dairies, horticulturists and breeders have over millennia rendered plants and animals into functional design objects. These organisms, and their modern GM descendants, are the objects that artist Richard Pell collects in his Center for Post-Natural History.[16] He argues that natural history museums collect the products of evolution, but human-designed species—from lab strains of *Escherichia coli* to modified goats—should be documented and preserved, too (figure 3.4).

Our nature is almost always unnatural: gardens, wheat fields, and pastures of grazing cows are human-manufactured constructions built from living parts; even wildernesses are state-managed parks. Clinging onto an idea of nature as separate from human activity is a futile and even damaging pursuit when it comes to thinking about what we design and its place in the world. "Nature" is just another human construct, intensified by design.

New Designs on Biology

We may have long designed *with* biology, but synthetic biology is proposing the design *of* biology.[17] This may not be just the iteration of nature: the selective breeding of plants or animals or another manufacturing revolution this time powered by biotechnology. It suggests a fundamental change in the things we consume. Biology becomes more than a material ripe for exploitation; it becomes both software and hardware for manufacturing; a toolbox for a new generation of designers mixing and matching components more akin to computer programs and components. The synthetic chemical age and its lurid plastics may have been an interlude as we return to an era of biological materials. How might synthetic biology's products fit into our classifications of nature and the designed products of culture?

In the 1970s, recombinant DNA—the ability to cut and paste genes from far-flung parts of the living kingdoms—sparked a still-continuing debate: Is this merely an extension of existing biological design or something new? Now, it is claimed that synthetic biology again offers a novel way to fashion biology more successfully into a tool for mass production, differentiated from the bespoke solutions of genetic engineering. Some academic researchers describe synthetic biology as revolutionary, which may help attract funding, whereas those in industry may prefer to call it an evolution, to keep it within existing regulation. I would argue that it is both. Although the technology builds on earlier ones, synthetic biology and its design of new systems and organisms presents novel dilemmas. New kinds of products, from rubber-producing microbes to bacterial computers, are prototypes for a different twenty-first century design, breaking our existing relationships with the things we consume. These designs may be unlike any we have previously known.

Biology is being remodeled into a design discipline in the name of progress, but progress and evolution follow different rules. Progress in technology is forward-looking, toward a future state of perfection. It also has a single, fixed-point perspective: that of the human. We even like to imagine ourselves as products of progress. Consider the linear improvement in the (incorrect) classic trope of human evolution, man striding off the page into the future, away from those hairy apes. Evolution, however, responds to context, not intention. Evolution connects all living things; as much as we impose our design on them, living machines, such as fuel-producing bacteria, are more loyal to evolution than human aspirations.

Nevertheless, technological progress and evolution can align. Darwin noted that "selection by humans should be understood within the context of natural selection."[18] Domesticated dogs bred for diverse human needs are still subject to the rules of natural selection. We humans are similarly co-evolving with our environment and technology and tools as described in chapter 1, further weakening the notion of the nature/culture divide. We may have got up off all fours and walked, but as long as we exist, we continue to evolve, too. As such, any products of synthetic biology will be intimately bound up in our own nature.

Nature is a human construct, and so too is the tree of life, the organizational tool we use to make sense of biology's diversity. The tree itself is always changing; its taxonomies are regularly reorganized and debated according to prevailing scientific understanding. Shifting from Linnaeus' two kingdoms in 1735 to Woese and colleagues' three domains in 1990, some experts even argue that the tree is "dead" and that life in all its varieties is better represented as a fuzzy ball. Certainly, the tree's simplicity masks nature's many complexities: agency, life, death, reproduction, combination, symbiosis, self-assembly, diversity, noise, context, emergent properties, and interaction with other living things. Biology is, ultimately, focused on survival. Such complexities are at odds with engineering ideals of control and simplicity.

As living things become design objects, we will have to consider the strategies design has developed to build its own successful role in consumerism: like function, form, desire, uniformity, obsolescence, and aesthetics. Questions that design has happily ignored become essential to examine, from life spans to a product's relationship with nature itself. Synthetic biologists propose technical design features—watermarks for identification, kill switches for self-destruction, or special guards to prevent horizontal gene transfer—to address the marriage of living things with designed products (examined further in chapter 6). If these new features are successfully integrated into biology, will it differ from the "natural" biology that already exists? Can we perceive living machines as either natural or unnatural, or do they demand a new category?

Synthetic biologists take a variety of approaches to make use of biology's diversity, defining design in different ways as they refactor, mix, digitize, and simplify it. The protocell—a biochemical machine assembled from scratch (the "bottom-up" approach)—is designed, perhaps more clearly than any other synthetic biological organism. But engineered bacteria modified from the "top down" are a more complicated prospect. Designed genetic circuitry is a mix of novel or redesigned DNA originally "copied" and "pasted" from other existing organisms. Once inserted into a naturally occurring biological chassis, the modified bacteria may vary only very slightly in terms of percentage change from wild types, but human design dominates the cell's function from our perspective. Self-assembling and self-reproducing, its progeny may not be crafted by human hands or human machines. But once the cell performs its designed function, the whole is labeled "designed"; a living machine is made. The redesign of the DNA code itself marks another approach. Jason Chin's lab at the University of Cambridge is one of several around the world seeking to invent a novel, parallel biology by developing an alternative code to DNA for biology to "run" on. Proponents of these "orthogonal" systems suggest that they may be easier to subject to human intention and to prevent from interacting undesirably with nature. "Orthogonal" systems may be biological, but they are products of human design.

Are these types of synthetic organisms any different from the life forms they once were or draw on? If they do diverge, where do we classify them within the tree of life? We may have to insert an extra branch into the tree to categorize them: a *Synthetic Kingdom* for designed and modified organisms that don't fit elsewhere (figure 3.5). *The Synthetic Kingdom* is an organizing device that mirrors synthetic biology's ideology, systematizing a new nature fashioned by engineering logic and its rationalization of the complexity of living systems.

When I first designed this extra branch, I saw it conceptually akin to an engineering solution to an engineering problem. It was intended as a tool to spark debate over our understanding of bioengineered organisms. It has proved useful: Scientists often comment to me that it is attached in

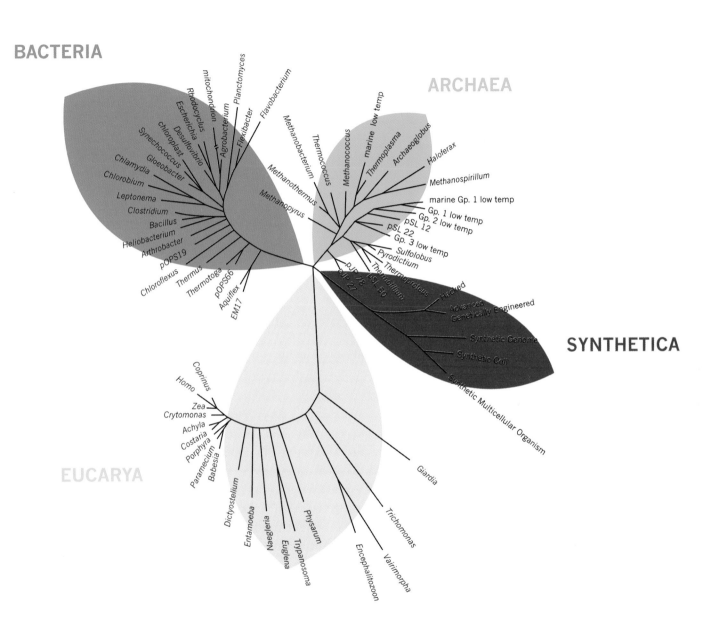

BACTERIA

ARCHAEA

SYNTHETICA

EUCARYA

Planctomyces
mitochondrion
Rhodocyclus
Escherichia
Desulfovibrio
chloroplast
Synechococcus
Gloeobacter
Chlamydia
Chlorobium
Leptonema
Clostridium
Bacillus
Heliobacterium
Arthrobacter
pOPS19
Chloroflexus
Thermus
Thermotoga
pOPS66
Aquiflex
EM17
Flexibacter
Agrobacterium
Flavobacterium

Methanobacterium
Methanothermus
Methanopyrus
Thermococcus
Methanococcus
marine low temp
Thermoplasma
Archaeoglobus
Haloferax
Methanospirillum
marine Gp. 1 low temp
Gp. 1 low temp
Gp. 2 low temp
pSL 12
pSL 22
Gp. 3 low temp
Sulfolobus
Pyrodictium
Thermoproteus
Thermofilium
pSL 50
pJP 78
pJP 27

Hacked
Advanced
Genetically Engineered
Synthetic Genome
Synthetic Cell
Synthetic Multicellular Organism

Coprinus
Homo
Zea
Crytomonas
Achyla
Costaria
Porphyra
Paramecium
Babesia
Dictyostelium
Entamoeba
Naegleria
Euglena
Trypanosoma
Physarum
Encephalitozoon
Vairimorpha
Trichomonas
Giardia

Figure 3.5
The Synthetic Kingdom, my
proposal for a new branch
of the tree of life to accommodate
our "new nature" (2009).

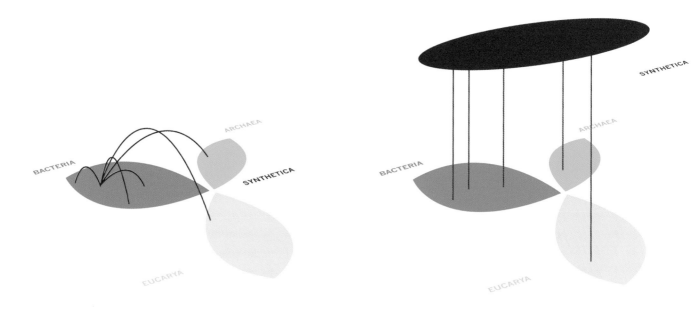

the wrong place. "It might be better placed coming out of a branch, not at the root," or, "How about it as a separate tree, or a cloud, or as networks of spaghetti," they say. Having such discussions about a fiction is illuminating: To me, it illustrates how inviting a suspension of disbelief helps us to imagine a different world view. That reasoned discussions prompted by a fiction can usefully address an issue is rewarding; the resulting iterations informed by these conversations help to raise new questions (figure 3.6).

Whether the branch should be smaller, differently placed, or more spaghetti-like, *The Synthetic Kingdom* itself has been viewed as veering between the critical and the celebratory. Have I given synthetic biology a kingdom of its own, effectively validating it and enforcing the separation between nature and culture for future products of synthetic biology? I see it differently: *The Synthetic Kingdom* puts our designs back into the complexity of nature, lessening the distinction between "our things" and "our selves." Acknowledging this connection between nature and what we design may allow us to design "better."

The modernist designers of the twentieth century argued that in terms of beauty, form followed function. For synthetic biology, the matter and meaning of designed things converges.[19] Our greatest challenge may be to acknowledge that the design rules for biology are unlike those for any other material. Human intention may not be enough to overcome evolution. Synthetic biology's designs on nature require us to adapt our understanding of design, the natural world, and life itself. With the prospect of change comes the opportunity to improve our thinking.

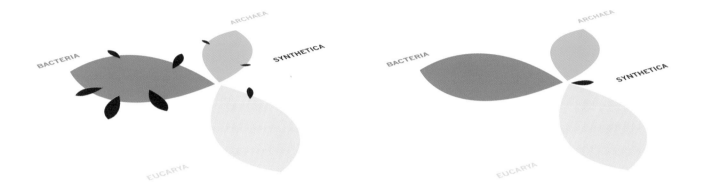

Machines for Living and Living Machines

Twentieth-century modernism advocated that the world could be improved through design. Modernist architect Le Corbusier saw the house as a "Machine for Living," a place where human efficiency could be enhanced through technological and functional design of the built environment. For Le Corbusier, humans were standardized parts in a rationalized urban system (figure 3.7). Modernism eschewed diversity, individuality, context, and bottom-up self-organization, prioritizing the architect's top-down control. Later, after these one-size-fits-all solutions were replicated around the world, it became apparent that the glass and concrete monoliths of the International Style did not solve all societies' problems. Cities from Africa to North America are still dealing with these insertions, as approaches to design shifted with the deconstructive backlash of the postmodern era.

Synthetic biology arguably follows these original modernist design cues. The J. Craig Venter Institute's synthetic life form, supposedly the first self-replicating species on Earth whose "parent is a computer," is widely described as a "living machine" (figure 3.8).[20] Corbusier and Venter's works are both objects described in the language of intentional design and control that celebrate the rational engineering paradigm and the merits of top-down design. While synthetic biologists may know less about the inner workings of the biological materials they use than did the engineers of Le Corbusier's concrete machines, both groups conceptualize design as separate from environmental or social context and arguably, the diversity and complexity of reality.

Figure 3.6
New iterations of *The Synthetic Kingdom*, drawn after discussions with experts. From left: the kingdom as a spaghetti network, a "floating" kingdom, small weeds on the existing branches, and a more "realistically sized" kingdom (2010–2011).

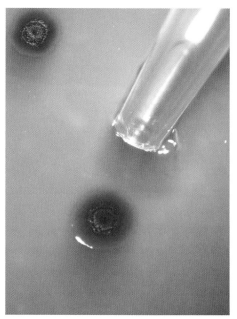

Forty years ago, biologist James F. Danielli, known for his work on synthesizing the first artificial cell from different components of an amoeba, highlighted this issue of a division between a technology and the societal factors surrounding it. Asked by *New Scientist* magazine in 1971 to share his predictions for synthesizing biology, Danielli outlined what was then just an imagined field of synthetic biology. His remarks remain curiously prescient, as does his concern for the ethical and social implications of such work.[21] Ending with a sober reflection, he described the societal burden of scientific progress:

> "But I agree that there are all sorts of borderline regions where things aren't so obvious. The trouble is that although vast sums of money are spent in science and technology in developing the research, only trivial amounts are spent on trying to predict the results of the work on society: as soon as something becomes available it is applied, without any study of what it might do to mankind." He looked wistfully out of his hotel window, the first trace of pessimism revealed. "If only we would spend at least as much money on studying the consequences of new technical discoveries as we spend on making them."[22]

What might be the consequences of technical discoveries that Danielli alludes to? Modernism was seen as the panacea for a rapidly urbanizing world, yet we now know that some of its most lauded theories for improving urban life had quite the opposite effect once built. And while living standards may have improved immeasurably for a proportion of the world's population, industrialization's success has brought with it unintended consequences that we now know have significant impact on the planet and our longer-term prospects. The unexpected outcomes of biotechnology need careful consideration, too. What is at stake, culturally, environmentally, and ethically as the materials of design come to life?

Appropriately, synthetic biology has been flooded with social scientists, bioethicists, and policy and risk experts, examining the potential promise and peril and evaluating whether the field raises novel issues. Investigation tends to focus on the same concerns: bioerror (the "right" technology going wrong, whether through design or user error, or the material deviating from human intention), bioterror (the "wrong" people using the "right" technology), and ownership (the technology lost in intellectual property thickets or monopolized). A fourth, potentially irresolvable category is the moral issue around designing life, or "playing God." In 2010, the U.S. Presidential Commission for the Study of Bioethical Issues deliberated on these issues and announced that effectively synthetic biology is no different in terms of risk and reward compared with previous technologies.[23] Their verdict: we should proceed, with caution. The underlying assumption is that all technology is good and vital for human progress.

Although Danielli's fantasy of a designed biology and the likelihood of a "synthetic future" may have become increasingly real, it is certainly not yet reality. But future products of the technology—and speculation on their impact—tend to be discussed in the many reports as if this is a fully realized technology. The scenarios that fuel the discussions swing between two poles of promise and peril, which Drew Endy wryly refers to as "the half-pipe of doom."

Many of the promises originate from within synthetic biology, its supporters constructing a one-size-fits-all utopia, marked by conflicting visions. Synthetic biologists describe a disruptive, sustainable technology that won't disrupt existing infrastructure, for example pumping out "green" jet fuel so that we can carry on as we are. Bacteria that will make materials and chemicals will be designed so that they survive only within the security of an industrial fermenter, yet the same species are also touted as potential powerful field technologies that will be safe to let loose to clean up toxic pollution. Synthetic organisms will work symbiotically with natural systems, nitrogen-fixing to improve crop production, but will not alter the ecosystem, thanks to kill switches. Open-source ideals will empower the developing world, but the technology is still being ring-fenced by first-world patent regimes to encourage investment. Friendly FBI agents liaise with do-it-yourself amateur biologists, assessing whether their work could ever present a potential risk, simultaneously raising the profile of "Do-It-Yourself" biology (DIYbio) while monitoring its activities.

These concepts are ultimately manifested in the twin spectre of the "dual-use dilemma," which describes a useful technology that can also be used to cause great harm (like an airplane or hammer). And synthetic biology certainly could well be both: the U.S. Defense Advanced Research Projects Agency (DARPA) is developing foundational technologies for manufacture called "Living Foundries" and "green" explosives, while the U.S. Defense Threat Reduction Agency (DTRA) sponsors the field's major conference series and wants to find ways to combat biological threat.

Building promise—and with it, hype—may be necessary for a new field of science to attract funding. But where do the dystopian extremes come from, if not just the active critics of the technology and DARPA/DTRA's involvement? If science sells utopias, the dialogue is set up in such a way that bioethicists called in to comment are often positioned toward the role of technological gatekeeper. They are asked to supply the "voice of reason," speculating on the potential impact of the technology, and in doing so may even invoke images of catastrophe to counter the utopian scenarios. Yet the arguments driving both sides are all too often made in the abstract and the extreme, plucked from a far-off future where the imagined technology is conclusively sophisticated. These are essentially tales of science fiction, and they lead to debate structured around world-saving green living or world-destroying tales of pandemic.

Certainly, it is not just government agencies and bioethicists considering the potential hazards of new biotechnologies. Speculating on the societal impact of new and imagined technology, often inspired by scientific discourse, science fiction novelists and screenwriters have long examined biological futures, frequently played out to catastrophic end. These include early classics like Mary Shelley's *Frankenstein* of 1818 or the monstrous animal hybrids of H.G. Wells's 1896 *The Island of Dr. Moreau*. By the 1950s, the decade of Watson and Crick's discovery of the structure of DNA, the implications of man-made biological life forms already anticipated the same themes of bioerror, bioterror, and ownership that still dominate policy and risk inquiry today. John Wyndham's intentionally designed, oil-producing, man-killing plants in *The Day of the Triffids* remain an evocative exemplar of bioterror.[24] Pohl and Kornbluth wrote of an overpopulated future in *The Space Merchants*,[25] where states exist only to support commercial enterprise, outlining the risks of a patented, technologized nature. Farm laborers tend monstrous protein blobs or work on vertical algae plantations in jungle skyscrapers. John Christopher's *The Death of Grass* depicts a natural virus decimating the world's major food crops.[26] Collapsing civilizations lie in its wake, echoing modern fears of pandemic caused by escaped or released synthetic organisms. Battling a stray, man-eating goo was Steve McQueen's leading-man debut in 1958; *The Blob* long presages Drexlerian concerns of self-replicating nanotech gray goo or synthetic biological green goo. With 60 years of scientific development behind us, we are still afraid of the same things.

Contemporary bio-fictions still unravel into spine-chilling dystopias, describing our world recolonized by nature, ruined by biology out of human control. Writer on landscape and nature Robert Macfarlane wonders if the underlying reason is a misanthropic slide back toward nature that helps us feel better about our postindustrial role in "ruining the world."[27] Nature will win again, eventually. These bio-apocalypses are often "cosy catastrophes," an accusation leveled at tales like Wyndham's where heroes survive unchanged in a world purged to a more natural state, simply shed of excess people.[28] Is this logic symptomatic of our entrenched view of humanity and its culture emancipated from the wilds? Nature is our enemy, constantly threatening us with both its nonliving and biological threats. The overuse of well-meaning, human-"invented" antibiotics triggers natural superbugs to kill us. Hubris over human intention and loss of control is easier to deal with than the unknown agency of the nature that we presume to master. These are the horrors that we transfer onto synthetic biology.

So how do we design for a world we want when that possible world doesn't yet exist, even in our imaginations? One of the difficulties for the development of policy and governance of synthetic biology is how to build a flexible and adaptive system that reflects current practice in the field but can also accommodate future developments. The balancing act between the safety of the "precautionary principle" often advocated by critics of the technology—treating something as dangerous until it can be proved harmless, which we may never be able to do—and desire for progress, is complicated.

As Danielli argued, we should be considering not only how or what we might design for this potential future, but what its effects might be. In 1990, the Human Genome Project, the multinational effort to sequence an entire human genome for the first time, invited examination of the societal issues that such knowledge of humanity might expose. The Ethical, Legal, Social Implications (ELSI) initiative was part of its remit. This work was placed downstream of technological development, implicitly stating that societal and environmental implications *follow* scientific development, rather than be part of it.

Vision may drive progress, and speculation can be useful. But viewing societal impact as an add-on is an error. I've heard synthetic biologists sigh, "If only they understood, they would want it!"—declaiming public misapprehension of genetic modification. But science and society are not distinct entities; scientists are part of society. The break between science and society mirrors the dichotomy between nature and culture, the psychological emancipation that stops us seeing our technology in the context of the ecosystem. Both are equally problematic as synthetic biologists attempt to design biology.

Working to improve the ELSI model, social scientists have experimented with moving the societal research "upstream" in synthetic biology, to earlier stages of the research process. While design too easily avoids responsibility for what it makes, these exemplary attempts within synthetic biology to include social considerations have been well meaning, but have on occasion become fraught.[29]

Despite these efforts, swinging speculations still dominate; the dystopias are shrugged off as "unbelievable," constructing a sense of inevitability to the direction of technological developments. As we rock between salvation and apocalypse, it is the more probable middle ground that proves harder to grasp. These are the closer-to-hand, incremental advances offered by synthetic biology that need analyzing now, which are less remarkable, as is the everyday life they represent. What will a world infused with biotechnology look, smell, or feel like? How will we have to change our behaviors and interactions and lifestyles? How do we protect ourselves from the corporate monopolization of living matter or make democratic decisions about appropriate levels of risk to the environment? It is the unexpected that often emerges from new technologies, rather than the neatly sign-posted paths of government plans. And there may be more than bioerror, bioterror, and intellectual property at stake.

While synthetic biology promises a better future through design, we should be wary as always to presume that there is one definition of "better," a one-size-fits-all future. Synthetic biologists' desire for standardization—whether minimal organisms or a universal biological "chassis"—threatens to standardize out the diversity and complexity of living things. A chassis may be "better" for maximal production of biofuel, but we should not assume that it means it is "better" for biology. If synthetic biology intends to design nature, its practitioners need to be fluent in the social and environmental issues embedded in their work: nature and culture, science and society are all interconnected. We need to design for multiplicity. But how do we design this bigger picture?

New Models for Biological Design

Design's engagement with our experience of the everyday means that it is a familiar language we can all effortlessly connect with. Working for industry, the designer's social responsibility has ultimately been economic. Design indeed can mean making beautiful "designed" things, but as Museum of Modern Art (MoMA) Senior Design Curator Paola Antonelli argues, design should be about making things meaningful.[30] "Designers stand between revolutions and everyday life," she asserts.[31] Rather than looking to existing design practice, it is emerging approaches in art and design that can provide useful models for a "better" biological design. Art and more experimental design practice can tease out problems, questions, and ideas not addressed by other disciplines, finding ways to express what we cannot yet put into words, including our fears. These new perspectives may help us negotiate the complex relationships between the living things that will be designed and the people they will be designed for, to help us think in more concrete terms about how a synthetic biological future might change us as individuals in particular, rather than the world in general.

At the fringes of design, revolutions in practice have existed since the Modernist era, as designers seek new definitions for their work, questioning embedded attitudes around design. This search for meaning reveals a desire to take more responsibility for design's role in contemporary culture, broadening its potential beyond the economic. This experimental design is where "innovation, functionality, aesthetics and a deep knowledge of the human condition combine to create outstanding artifacts."[32] Although this kind of design may still be concerned with function and utility, its interest in imperfection, rather than uniformity, is more aligned with art. Architects Jean-Gilles Décosterd and Philippe Rahm take this approach with their investigations into "physiological architectures." Décosterd & Rahm's ephemeral built spaces are not defined by their physical limits, but by their inhabitant's physiological response to stimulation. By manipulating light levels or the air's chemical composition, they challenge our conventional understanding of architecture as a physical structure, redefining it as the relationship between our body's biochemistry and the space we inhabit (figure 3.9). As design borrows and even blends into art (and scientific) practice, it can be used to investigate not only new functional potential but also philosophical and aesthetic issues raised by new materials and experiences.

Designers are increasingly realizing the societal impact in using design as a medium to trigger debate and discussion. British designers Anthony Dunne and Fiona Raby have been instrumental in developing "critical design" or "design for debate." Their *a/b* manifesto describes this shifting role of designer from problem-solver to philosophical sense-maker (figure 3.10). Designers can become "problem finders," Dunne and Raby suggest, identifying glitches in the system. Antonelli proposes that by asking, "What is this about and can we do something about it?" designers can go further, becoming "problem

Figure 3.9
Décosterd & Rahm's *Diurnisme*
(2007) installation makes "night
during the continuous artificial
day of modernity." The installation
at the Musée National d'Art
Moderne, Centre Pompidou,
uses bright orange-yellow light
with wavelengths above 570
nanometers, which are perceived
by the body's clock as "true" night.

makers." By seeking out questions, we can reveal new perspectives on the world; as problem makers, we can challenge existing systems and design ways we might change them.

These attitudes can cross into synthetic biology's design discourse. While Dunne & Raby "design for designers" to stimulate new modes of practice, designing for synthetic biologists may well be similarly useful.[33] Moving away from designing goods to consume, to designing ideas and questions, designers have a potential role upstream in science, where new technologies are made. This "design without commerce" holds with Bruno Latour's notion of an object-oriented democracy, making, as design historian David Crowley suggests, "a forum where science and its objects are put under public scrutiny."[34]

(a)	(b)
affirmative	critical
problem solving	problem finding
design as process	design as medium
provides answers	asks questions
in the service of industry	in the service of society
for how the world is	for how the world could be
science fiction	social fiction
futures	parallel worlds
fictional functions	functional fictions
change the world to suit us	change us to suit the world
narratives of production	narratives of consumption
anti–art	applied art
research for design	research through design
applications	implications
design for production	design for debate
fun	satire
concept design	conceptual design
consumer	citizen
user	person
training	education
makes us buy	makes us think
innovation	provocation
ergonomics	rhetoric

Despite calling for change, Antonelli believes designers should remain generalists: "Designers should keep on doing what they do best, which is to address... the sensible, human and beautiful production of things in the world."[35] Decades ago, Buckminster Fuller anticipated a "comprehensive designer." For him, "The specialist in comprehensive design is an emerging synthesis of artist, inventor, mechanic, objective economist and evolutionary strategist."[36] This might still ring true. Future designers may need to understand biology, but they also need to know the world it operates in. Artist, computer scientist, engineer, and experimental designer Natalie Jeremijenko works in this space, unfazed by changes in scale or material between the lab, the studio, the community, and the ecosystem. Her *OneTrees: An Information Environment* design research project appropriates the tree as a biosensor. In 2003, 1,000 genetically identical cloned specimens were planted throughout the San Francisco Bay Area as a network of environmental biosensors and long-term record-makers.[37]

But synthetic biologists—designers of biology—and designers are not one and the same. The designer's flexibility to adapt to different roles and act as a go-between for experts and nonexperts offers a complementary set of skills to the knowledge of biology's complexity required of the synthetic

Figure 3.10
Dunne & Raby's *a/b: a changing understanding of design*, a manifesto for new thinking about design (2009).

biologist. An emerging role for the designer is a form of social critic. Here, the designer acts as a mediator between science and public interest and its social desires, addressing the fact that science is "concerned with what is, not with what ought or ought not to be" further down the line.[38] This sounds a little like the public intellectual, a role in decline. "Intellectuals in different guises play a crucial role in initiating dialogue and engaging the curiosity and passion of the public," sociologist Frank Furedi writes.[39] While this can manifest as public engagement with science, more interestingly, it can face the other way, as scientific engagement with the public, or better, work to eliminate the divisions between the two. This has been the direction of an evolving critical design, as some of its practitioners—many taught by Dunne & Raby (including myself)—develop a type of applied, speculative bioethics. Researching science and engaging in discussion with scientists, these designers speculate on future interactions with biotechnology by designing tangible objects, developing new visual metaphorical languages in the process.

One such project is David Benque's *Acoustic Botany*, which challenges synthetic biology's promise to design for our needs (figure 3.11). Instead, he wonders what it would be like if it were designed to feed our most irrational desires, like many other consumer products today. Designing in dialogue with scientists, he imagined a *Genetically Engineered Sound Garden*, designed purely for aesthetic pleasure. In his fiction, varietals are engineered to produce different sounds, some even relying on symbiotic relationships with insects to "pluck" their fibers or with modified agrobacteria, producing air for the instruments.

Some philosophers of science suggest that science uses models as useful fictions to explore the world.[40] The model helps to test a hypothesis. If wrong, the hypothesis (and the model) can be rejected, and science continues with this knowledge noted. Useful fictions could quite literally help us explore possible worlds: Robert Macfarlane asks whether, instead of nostalgically looking backward to a nature untainted by human hubris as described in the previous section, can we design our fictions as "constructive ruins" that help us better imagine our own present and "choose more wisely between our own available futures?"[41] These "complicatedly forward-looking" ruins could help us feel "crisis in our guts" before we progress. Testing these fictions by experiencing them, we examine our personal responses, socially, politically, ethically, and culturally. Combining these two very different concepts of the fiction, from philosophy of science and literature, could help us to imagine how we could think about applications and implications holistically from the outset.

Useful "design fictions" or "speculative fictions," such as Benque's *Acoustic Botany*, model potential worlds. They are not intended to be predictive, the preserve of futurists who imagine what technologies might come

next; rather they test and explore the everyday interactions of much nearer possible futures. They aim to confront complex societal issues by imagining unrecognized things or situations, critically commenting on—or triggering for debate—behavior and fears through our reactions to objects and our use or misuse of them.

Useful fictions in science may be built on the proposition, "as if . . .,"[42] but this work often sits in the realm of "what if . . .," the eponymous title of a series of exhibitions curated by Dunne & Raby, showcasing the work of their students (including myself) in this vein.[43]

The *Microbial Kitchen*, a "Philips Design Probe," is a design fiction with the gloss of corporate design language and is intended to test, not predict, how we might live with biotechnology in our homes (figure 3.12).[44] But produced and researched within a corporate setting, its designers may well influence the way that future biotechnologies are developed. Scientists admit to being inspired by science fiction on occasion. This may be due to the authors' thorough research into current science and its likely trajectory, rather than their stories actually influencing research directions. But

Figure 3.11
Lab Testing Rig from designer David Benque's *Acoustic Botany* (2010), a design fiction about plants engineered to produce sound. Factors like tension and temperature are modulated to fine-tune the sound. Benque asks if we might engineer biology for reasons of pleasure, not just need.

as I have learned through my own experiments collaborating with synthetic biologists: in imagining the future, you may make it more likely. Useful fictions can unintentionally become embedded in the language of the field, even shaping it. These are the well-informed ideas that straddle a delicate boundary, so well rendered in the accessible consumer language of design, they become difficult to identify as fictions.

If they look too real, speculative fictions risk losing their critical edge. Issues that the designer wants to be debated simply become acceptable, the power of "what if . . ." deflated as the audience is desensitized to something previously quite unfamiliar. Rather than testing possible worlds, the fiction becomes a kind of sophisticated endorsement of a now-more-probable future. An example of this in a collaborative project of my own, *E. chromi*, described in chapter 6. Finding ways to design and test futures without prioritizing them and preventing other outcomes is important for designers or artists working with scientists in this way.

"Bio art" works differently to investigate biotechnology's impact. Bio art is an emerging branch of contemporary art "that manipulates the processes of life" as bio artist Eduardo Kac explains. It uses biological materials or organisms, as well as biotechnology's tools and processes, often toward "unusual or subversive" means.[45] Bio art is a way to initiate discourse over science and often brings scientists and artists to work together. But many bio artists are unhappy with their work with living materials being co-opted as a descriptor of technologies and a tool for public engagement. One solution for these artists is to work with scientists, but separate their work conceptually to avoid diminishing their role as provocateurs. As Oron Catts, of the Tissue Culture & Art Project and SymbioticA (and Synthetic Aesthetics resident) explains, in this disciplinary integrity and avoidance of making "useful" things lies the ability to create shock to create discourse. Yet both Kac's infamous green fluorescent *GFP Bunny* from 2000 or Tissue Culture & Art's *Semi-Living Steak*—a slab of ironically titled "victimless meat" grown, fried, and eaten in a gallery in 2003 as *Disembodied Cuisine*, described in chapter 2—have over time encountered the strange frontier where provocation becomes absorbed into and even part of scientific progress. "Victimless meat" has become a stock phrase in media reports on the burgeoning industry of lab-grown burgers.

This awkward line shows how neither art nor design is immune from instrumentalization by the science it works to critique. These examples also demonstrate how important it is to continue to interrogate the cutting edge. It is through the *Semi-Living Steak* project, not the recurrent media hype about a future of plentiful lab-grown meat, that we are reminded that victimless meat is no such thing, fed as it is on fetal calf blood products, and we are obliged to consider the strange "semi-living" status of tissue culture.

Synthetic biology not only presents problems that need to be solved but also presents dilemmas. Designers and artists can work as "provocateurs," seeking out and testing these predicaments. Working within science rather than separate from it is a way to ensure such investigations are informed by scientific developments and that they are considered by the scientific community, removing the imagined divide between science and society. The different models of contemporary design and art practice described here illustrate ways that designers and artists are working to challenge the accepted boundaries of their own disciplines. These approaches could inspire new, more collaborative practices between synthetic biology and social science, art, and design that help to open up the discussion of what could or ought to be, not just what is, questioning the assumptions held.

As Part Three documents, the Synthetic Aesthetics residencies have followed this route in their collaborations. Some of the projects suggest a synthesis of approaches between synthetic biology, art, and design, hinting at "critically-engaged biological design." Where biological things were physically made, they were real, like Christina and Sissel's human cheese (chapter 17), or Oron and Hideo's circuit board digesting-algae (chapter 11), more like the "real" products of bio art than the props of critical design, avoiding the possibility of being mistaken as fictions. While these objects were real, they were functional, too, countering the anti-utilitarian ethos of bio art. Their strength is in the way that they are useful: they are provocative, challenging core assumptions embedded in the technology. These living organisms, tools, and ideas provide new insights into synthetic biology. Useful fictions as models of possible futures need not be science fiction, but can be experimental artifacts.

Synthetic biology suggests a different nature, and a different world; we need to think about what we want from this biological future. As synthetic biology attempts to design a new biology, there is an opportunity to reinvent design: If design is engaged in these technologies, it should proactively claim a role in shaping them, too. Design (and art) can bring tools for useful critique, debate, collaboration, and investigation into science, while bringing the tools of science to the expertise of others in society. For synthetic biology to be a successful future design practice, it should consider applications and implications together, a mode that emerging design practice is investigating. As a discipline, synthetic biology should include not just scientists and engineers but also artists and designers, social scientists, and risk and policy experts. This is not a public engagement process or a search for predictions; it should be a critical part of the scientific process, if we hope to design a new nature well. Here lies the opportunity to design a disruptive technology, one that might actually challenge entrenched modes of living and consumption and challenge the prevailing attitude that what we make is somehow separate from the natural world.

4.
Nature Is Designed

Drew Endy

I am the family face;
Flesh perishes I live on,
Projecting trait and trace
Through time to times anon,
And leaping from place to place
Over oblivion.

  —Thomas Hardy

Earth is dancing with life. What we perceive to be natural biological systems exist across so many distinct and diverse terrestrial environments and appear to have done so for billions of years. We still do not know exactly how life started here or arrived from somewhere else. Most of us expect that life will continue for quite a while still. As humans, individually and collectively, we can profoundly affect the biosphere, from single organisms to entire ecosystems. Yet we do not control all of life and likely never will.

  To restate the above conclusion in the context of synthetic biology, note that there are at least 10^{35} base pairs of natural DNA on our planet today, spontaneously reproducing and evolving moment after moment. Meanwhile,

Frontispiece
Nature Is Designed.

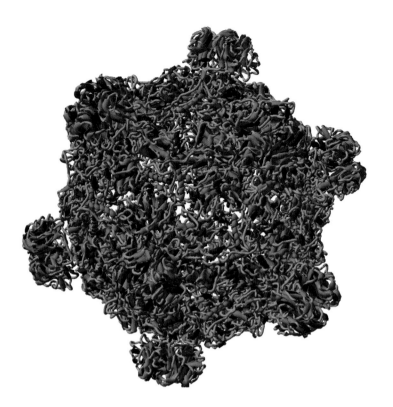

Figure 4.1
Atomic representation of
the proteins comprising a
bacteriophage øX174 particle.

we can likely now only synthesize 10^{11} base pairs of DNA per year. Even if our DNA synthesis capacities double every year for the next 80 years,[1] and no other factors became limiting, only then we would start to match the DNA construction capacity already on our planet and beyond our control. However, even if we could make so much DNA, we would likely not know how to design at such an immense scale.

So, how can we possibly think about design and biology, and what does this have to do with synthetic biology? Certainly, we have made a few engineered biological systems and will make many more still, but most of these organisms are or will remain contained artifacts meekly existing in the context of a grander natural process. This chapter instead explores two possibilities, one strangely alien and another well accepted. First, perhaps someone else designed life already: What we consider to be natural living systems are actually another's designs. Second, that the natural processes through which life continues to exist spontaneously generate constrained systems that are impressively optimized: While we imperfectly understand such natural evolutionary processes, we will become increasingly responsible for their design.

The most numerous and common reproducing biological systems that we know about devour bacteria. These "bacteriophages" are readily found wherever microbes exist. Each bacterial cell infected with a phage can produce hundreds to thousands of progeny virus particles, often starting within just a few minutes of infection. For example, the photosynthetic microbes that exist in the surface waters of Earth's oceans are preyed upon by bacteriophages to such an extent that marine microbiologists now estimate[2] that there are 10 billion bacteriophage particles per liter of ocean surface water (figure 4.1). Think about that the next time you go for a swim.

Perhaps because phage particles are all over, and also because they are small and easy to grow compared to most other natural biological systems, bacteriophages were some of the first biological systems ever collected and studied in detail more than a century ago by the scientists who launched what became molecular biology and genetics. Now with synthetic biology, phages continue to enable pioneering work, from the first rationally redesigned genomes to the provision of the molecular components needed to implement engineered living computers.

But what can we say about the designs of the bacteriophages themselves? One fantastic possibility became apparent in the 1970s with the invention and application of DNA sequencing technology. Today, DNA sequencers are well known as machines into which physical DNA molecules are added and from which each DNA molecule's atomic code is read out as a string of information represented by four letters (e.g., TAATACGACTCACTATAGGGAGA). However, with the invention of DNA sequencing, something's genome had to be sequenced first. It turns out that the first DNA genome sequenced encoded a bacteriophage that scientists had collected and placed in a test tube in the 1920s. The tube was labeled "øX174."

Sequencing of the øX174 (pronounced "phi X one seventy four") genome quickly confirmed much of what had been hypothesized about DNA,[3] such as proteins being encoded by stretches of DNA (genes) that start and stop at defined sequences (figure 4.2). But the full sequence of the øX174 genome also greatly strengthened a funny question that had been developing. Specifically, does a given stretch of DNA only encode a single gene that specifies just one specific protein? Would George Beadle and Edward Tatum's "one gene, one enzyme" hypothesis from the early days of molecular biology hold true? Apparently not.

Instead, the øX174 genome sequence provided irrefutable evidence of genes inside other genes, including nested genes that code for totally different biochemical functions. For example, the region of øX174 DNA coding for gene A also encodes the smaller gene B. The gene A sequence provides the code for a protein that helps the virus to replicate its DNA during infection. The gene B sequence codes for a scaffolding protein that helps assemble the virus particle—totally different genes written in overlapping form on the same stretch of DNA.

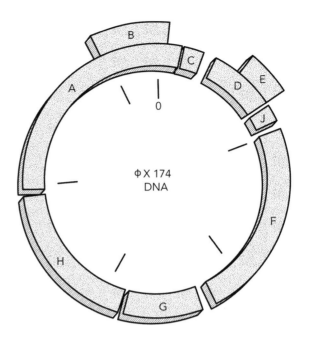

How can one gene be nested within the coding sequence for another? Where do genes within genes come from? An answer to this first question can be figured out given basic knowledge of how the genetic code works. We will return to the second question and the code itself later, but, for now, note that natural sequences of DNA that encode proteins use what is known as a "triplet code," in which three bases of DNA code for one amino acid within a protein. This triplet code is necessary because natural DNA only uses four letters (A, T, C, G) at each position, but proteins can be made from 20 distinct amino acids. A "singlet code" would only allow one of four amino acids to be chosen. A "doublet code" could determine one of 16 (4 × 4) amino acids. A "triplet code" could handle up to 64 amino acids (4 × 4 × 4), a "quadruplet code" 256 (4 × 4 x 4 x 4), and so on. Given 20 amino acids, three is the minimum (and magic) number. Knowing all this, DNA coding can get interesting. Let's say I have a stretch of triplet DNA such as ATG CTG CAT GAC GAC CGA CTA GCC. Each three bases of DNA code for a particular amino acid, read from the first ATG triplet, here followed by a CTG that codes for the amino acid leucine. But what if we changed how we looked at the DNA just by shifting the boundaries of each triplet a single position? The same DNA sequence now reads differently: A TGC TGC ATG ACG ACC GAC TAG CC. Note that a second ATG triplet appears inside the original sequence followed by an ACG that codes for a different amino acid, threonine. Such "frame shifting" of triplet-coded DNA actually allows for up to three totally

Figure 4.2
Schema of the bacteriophage
øX174 genome; lettered boxes
indicate different genes.

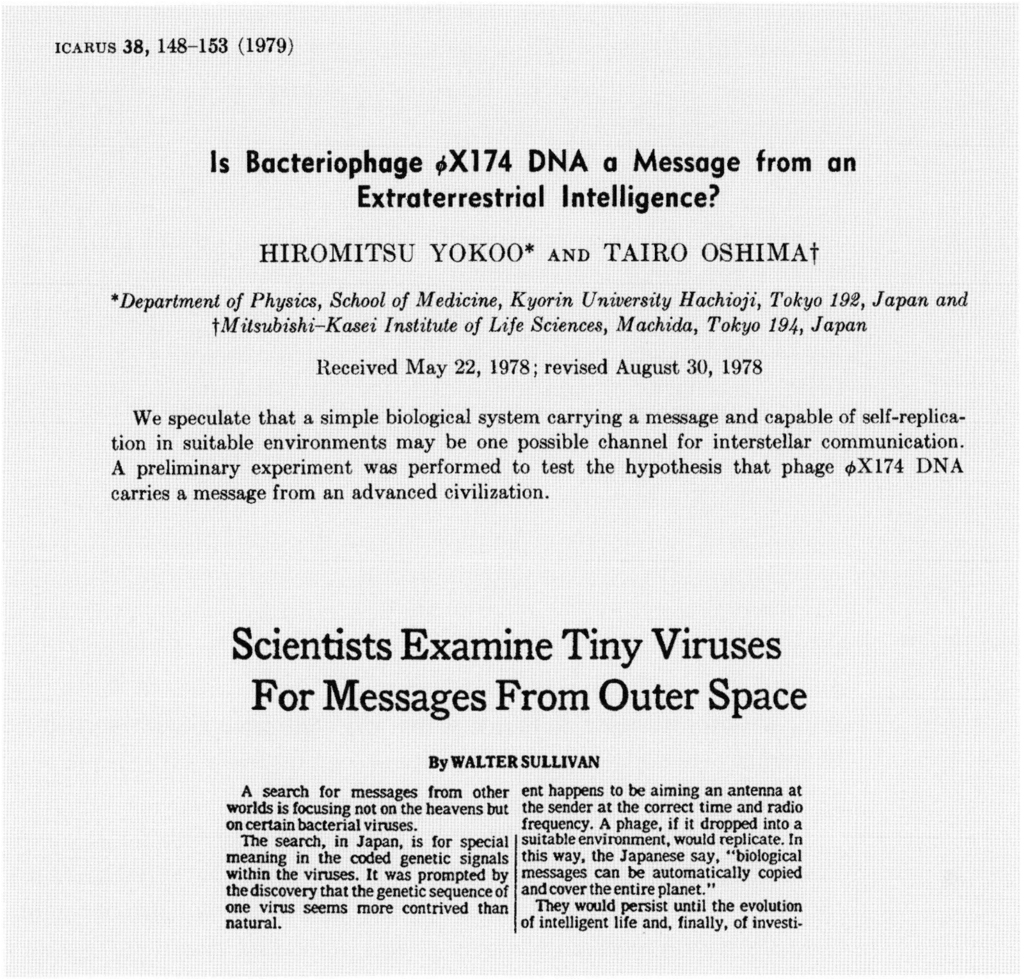

ICARUS 38, 148–153 (1979)

Is Bacteriophage φX174 DNA a Message from an Extraterrestrial Intelligence?

HIROMITSU YOKOO* AND TAIRO OSHIMA†

*Department of Physics, School of Medicine, Kyorin University Hachioji, Tokyo 192, Japan and
†Mitsubishi–Kasei Institute of Life Sciences, Machida, Tokyo 194, Japan

Received May 22, 1978; revised August 30, 1978

We speculate that a simple biological system carrying a message and capable of self-replication in suitable environments may be one possible channel for interstellar communication. A preliminary experiment was performed to test the hypothesis that phage φX174 DNA carries a message from an advanced civilization.

Scientists Examine Tiny Viruses For Messages From Outer Space

By WALTER SULLIVAN

A search for messages from other worlds is focusing not on the heavens but on certain bacterial viruses.

The search, in Japan, is for special meaning in the coded genetic signals within the viruses. It was prompted by the discovery that the genetic sequence of one virus seems more contrived than natural.

ent happens to be aiming an antenna at the sender at the correct time and radio frequency. A phage, if it dropped into a suitable environment, would replicate. In this way, the Japanese say, "biological messages can be automatically copied and cover the entire planet."

They would persist until the evolution of intelligent life and, finally, of investi-

different proteins to be programmed as overlapping genes within any given DNA sequence!

So we know that it is possible for genes to be encoded within other genes. But, how could such a genetic architecture evolve? Where do nested genes come from? As a researcher, I can relate lots of explanatory stories, most of which are based on indirect evidence, but I still do not fully know. Stated differently, I could not now teach you to design DNA sequences in which genes start out separated and then later, as an organism reproduces and evolves, spontaneously begin to overlap. While I could guess at a lesson plan, I have never seen a report detailing direct observation of a natural gene overlapping process.

However, back in 1977 and given the full øX174 genome sequence, researchers in Japan had a totally different idea. Perhaps genes overlap in øX174 because somebody or something intentionally designed them to

Figure 4.3
Alien messages in øX174 gene B? Headlines from the Japanese research paper and the *New York Times*.

overlap! Going further, given that there was no evidence of past advanced civilizations on Earth with a capacity to design life, the Japanese researchers hypothesized that such DNA sequences would need to be designed by an extraterrestrial intelligence (figure 4.3).

Now, before you dismiss their idea as being totally crazy, consider two additional facts. First, the length of DNA coding for gene B, the small gene hidden inside A, is 121 triplets, which is the square of a prime number (11 × 11). Perhaps each triplet specifies one part of a hidden message that can be read out only when displayed on a square grid? Perhaps the author(s) of the hidden message used prime numbers to provide a clue for where in the øX174 genome we should look for the message? Second, just a few years earlier, Francis Crick, one of the co-discovers of the double helical structure of DNA, co-authored a paper exploring the idea of "directed panspermia";[4] specifically, that an alien civilization could send life throughout the universe by transmitting small payloads of microbes (and presumably viruses of microbes) between solar systems. Because the universe is ~8 billion years older than when life is thought to have first existed on Earth, there would have been plenty of time even for very slow spaceships to make such a journey and seed new planets, including ours. And, if you want more to think about, what's the chance that the first DNA genome ever sequenced by humans would encode nested genes of lengths involving the squares of prime numbers?

Sadly, it turns out that Hiromitsu Yokoo and Tairo Oshima, the Japanese team, concluded that they could not find any hidden messages in the øX174 genome. This is too bad for us because if all life on Earth had arisen from purposeful molecular designs by others, then our questions regarding the relationships between biology and design would become much simpler, philosophically speaking. Perhaps someone might find such evidence in the future. However, I do not know of a single researcher who is now scouring the information encoded within natural DNA sequences for hidden extraterrestrial messages, which seems strange to me given all the interest in listening for interstellar radio communications.

Thus, for now, we must return to the second possibility regarding biology and design. Life started on Earth a few billion years ago and has since continued to exist. This process involves individual organisms that reproduce alone or with one or more partners. Offspring can become slightly different from their parents and siblings via spontaneous changes that arise during reproduction. These differences lead to diverse populations in which individuals compete against one another and also against other types of organisms. Competition can be direct (e.g., predator eating prey) or indirect (e.g., number of offspring). Certain types of organisms and their specific features provide

competitive advantages (e.g., a horse-like creature with a slightly longer neck in an environment wherein most available food is leaves at the tops of tall trees). Individuals with favorable traits tend to have more offspring, who tend to have features that are similar to their parents, and so on, from one generation to the next. Over time, these features take on the appearance of designs, in which individual organisms develop highly refined abilities enabling survival and eventually reproduction, thereby transmitting their features from one generation to the next (e.g., Thomas Hardy's "family face" carrying trait and trace over oblivion).

Figure 4.4
A fly's eye.

As one example, consider the eye of a fly (figure 4.4). Being able to see helps a fly to find resources, mate, and avoid fly swatters and other hazards. How good are flies at seeing? It turns out that the short answer is "amazingly awesome." Flies are now somehow so good at seeing that it would be practically impossible to design a better visual system for flies. Bill Bialek, a biophysicist at Princeton, introduces the situation as follows: "We have found that computation in the fly's visual system works with such precision that it is limited by photon shot noise and diffraction, that information transmission across synapses between neurons operates near the limits imposed by the "quantization" of signals into discrete packets of chemical transmitter, that deep inside the brain signals are represented by sequences of action potentials or "spikes" with nearly optimal efficiency, and that these levels of performance are the result of dynamic adaptation processes that allow the system to adjust its strategies to the statistics of the current visual environment."[5] Stated differently, the eyes and brain of a fly can observe things almost perfectly well, very near the physical performance limits of its visual system. Does this constitute an example of biological design? Yes, of sorts. Albeit a "design" that is believed to have evolved over time as individual flies that could see a bit better than their siblings produced greater numbers of better-seeing offspring.

Are these types of evolved biological designs only found in the macroscopic world that we are familiar with—the physical arenas in which individuals compete? The fly's eye, the giraffe's neck, our opposable thumbs and relatively large brains all seem readily reasoned as systems that we might expect to become highly optimized via evolutionary processes. But what of the underlying molecular architecture that life itself is composed of? Does "design" make any sense when we consider biology at the level of atoms and molecules?

Let's start exploring the potential for biological design at the atomic level by considering the details of deoxyribonucleic acid, or DNA (figure 4.5). Most natural living systems encode heritable information within DNA molecules. What is meant by heritable information? As mentioned in discussing øX174, changing the "letter" or base of DNA at any given position within a gene can change the amino acid incorporated into a protein. In turn, changing even one amino acid within a protein can sometimes have significant effects. For example, replacing a single triplet that codes for glutamine with another that instead directs valine to be incorporated into hemoglobin, the oxygen transfer protein found in red blood cells, creates "mutant" hemoglobin molecules that stick together, deforming red blood cells and hindering the flow of blood through capillaries. It turns out that if you have one copy of the "mutant" hemoglobin gene, you are more likely

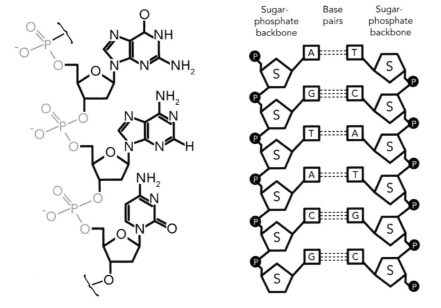

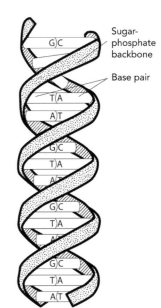

Sugar-phosphate backbone

Base pair

to be resistant to malaria, a disease in which a parasite infects red blood cells. However, if you have two copies, you will suffer from the condition commonly known as sickle cell anemia. Whether you have zero, one, or two copies of mutant hemoglobin depends on either the heritable information stored in the DNA sequences of your parents and passed on to you or that a random mistake during the copying of what became your DNA produced this hemoglobin mutation spontaneously.

So, could we say that DNA is designed to store information? Or, more carefully, that the performance of DNA in storing heritable information is amazingly awesome, like the eye of fly, so much so that we would be hard pressed to design something totally different that was somehow better? To consider these questions, we need to study how individual atoms are arranged in DNA molecules and how the atomic arrangement itself allows for information storage. Typically, when a researcher describes any particular sequence of DNA, he or she talks about "bases" or "base pairs." We'll return to this aspect of DNA in a moment. For now, focus on the "boring" part of DNA, which is known as the "backbone." Think of a single DNA molecule's backbone as a string made from a repeating pattern of sugar molecules linked together by phosphate groups. The sugar is ribose, each molecule of which contains five carbon atoms arranged in a pentagonal ring. Each phosphate group includes a central phosphorus atom surrounded by four oxygen atoms. One key fact to note is that phosphate groups have an extra electron and thus a negative ionic charge.

Figure 4.5
Atomic chemical structure of single-stranded DNA, showing the negatively charged phosphate backbone (left), base pairing among double-stranded DNA (middle), and the DNA double helix (right).

If you remember the adage "opposites attract," then consider its partner, "like repels like." Now, look at the DNA molecule backbone made with a string of negative charges at each phosphate group. Like repels like. If a single DNA string starts to fold back upon itself, then the negatively charged phosphate groups will repel one another, keeping the DNA molecule straight for some length. Steven Benner confirmed this property when his team of chemists synthesized a variant of DNA that has a neutral (i.e., no charge) backbone, finding that their neutral molecules didn't work well for storing biological information.[6] Why?

To answer this question let's return to the "bases" of DNA. These are the atomic groupings that specify the information encoded within DNA, such as what type of hemoglobin protein you make. Different arrangements of atoms make up the four bases, commonly represented by their abbreviations (A, T, C, G). To be useful as an information storage system, a DNA molecule must be able to replace any one base with another without having the entire DNA molecule change its chemical function. It is very strange for a molecule to not change its properties when you start moving atoms around. For example, water (H_2O) becomes hydrogen peroxide (H_2O_2) just by adding a single oxygen atom. How can DNA exchange many atoms at a time (one base for another) yet maintain its overall properties?

Returning to the work of Benner and others before, it seems that the negative charges produce a "backbone" for DNA in more than name alone. Changing one base to another does not override the fact that the phosphate groups cannot get too close without repelling one other. Thus, regardless of the specific sequence of bases, a DNA molecule will tend to posture (i.e., stick out) its bases for pairing with the bases of another DNA molecule (hence, "base pairing"). Changing a base via mutation or otherwise does not typically change a DNA molecule's fundamental structure or chemical activity, rather only the information stored within. Of note, this specific characteristic of DNA seems absolutely essential in order to support evolution. Without an ability to change bases with chemical impunity, the individual genetic mutations that arise during evolution (e.g., in hemoglobin and other genes) might otherwise destroy the overall capacity of DNA molecules to function.

Many additional stories of natural living systems solving problems by arranging and controlling atoms, cells, tissues, and organisms have been worked out. As a lead-in to just one more story, the pattern relating each DNA triplet to one of the 20 amino acids (i.e., the "genetic code") appears itself to be optimized in support of evolution (figure 4.6). Random changes in DNA coding for a gene tend to produce evolutionary-favored changes in the amino acids composing a protein (when compared to changes arising from most randomly selected genetic codes). For now, there seems no end

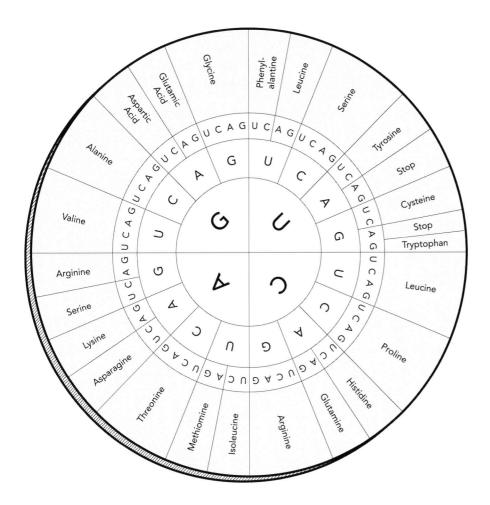

to the layers by which biological systems have been highly optimized to solve physical problems in order to continue to exist. To an engineer learning of this emerging reality, it feels simultaneously amazing and intimidating. Could we ever design biology?

Answering this question today, given the incredibly young and immature state of biotechnology and its overwhelming focus on immediate health and bulk manufacturing applications, seems premature. Moreover, most engineers and scientists working in synthetic biology do not now intend to design new living systems that directly compete on the basis of their raw physical performance against natural evolved systems. Instead, natural systems are being reworked through breeding or engineering for new purposes. Some of these activities have and will continue to involve aesthetic choices (e.g., dog breeds). Others focus on utilitarian outcomes (e.g., a yeast engineered to make a therapeutic drug). There certainly are modes of design in such honorable work, but it is not what we have explored here.

Figure 4.6
Nature's conventional genetic code, by which specific DNA triplets (three inner rings) encode one of twenty amino acids (outer ring).

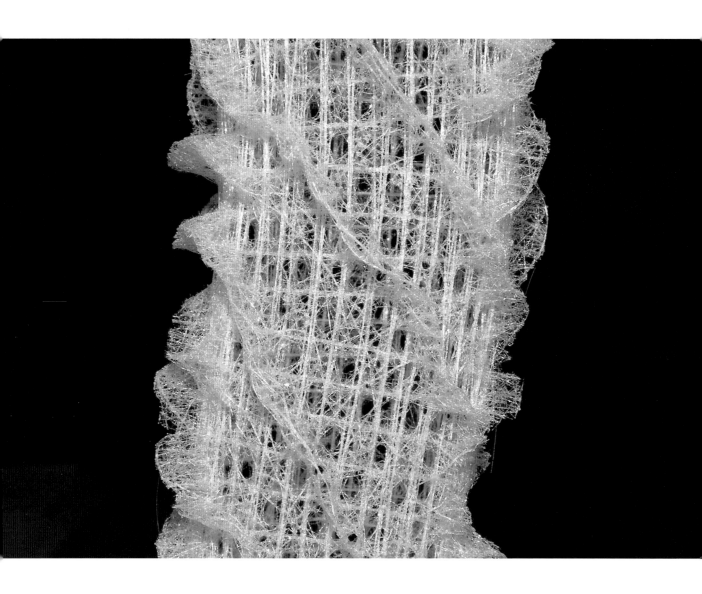

Figure 4.7
The Venus flower basket sponge builds an elaborate woven skeleton of fine biosilica strands using the minerals in seawater—an impressive example of biological design and manufacturing.

Instead, imagine that you are walking along a beach. Ask yourself, could you design a computer from what you see about? From chapter 1, you are likely to recall that modern computing chips are engineered from silicon, and perhaps also note that the sand you are walking on is a form of silicon, too. Consider all the changes to sand that are needed to make a silicon microprocessor: refinement to a purer elemental form, growth of a crystalline wafer, a variety of masks and etchings along with deposition of dopant atoms, just for a start. When we gain a capacity to rework biological molecules and materials to such an extent like what we now routinely realize with silicon and other natural substances, then I expect we can start to realize a new type of biological design (figure 4.7).

Returning to the beach, could you imagine designing a totally different type of computer from what is available to you? What about making a computer from all the living materials? For example, could you design a computer from the bacteriophages that surf within the waves? What if you took the proteins that phages sometimes use to silently embed their genomes within their hosts and instead made genetic data storage systems[7] and perhaps Boolean logic gates?[8] What if you hijacked and repurposed the phage particles themselves to transmit DNA encoding your genetic memory and logic elements among cells?[9] Instead of replicating mindlessly, would your engineered cell–cell communication phage turn a population of microbes into a self-mixing goopy computer? When we can fully refine and repurpose all of biology to make things that are as distant from sand as a silicon computer chip, we will be liberated from direct comparisons to natural living systems and their practical evolutionary constraints. Only then might we realize an opportunity to explore a new mode of biological design.

5.
There Is No Design in Nature

Pablo Schyfter

There is no design in nature. People, *and only people*, design. Nature—here specifically biological nature—does not design. As this chapter demonstrates, a number of things distinguish what happens through the different processes of evolution and what people do when they design something. I examine many differences but ultimately focus on just one: values. When people design, they base their design choices on norms, mores, principles, and so on. In this sense, synthetic biology's aim of "designing with nature" is one further project in bringing to bear our values onto the shape of the living world. It becomes imperative to ask what we have failed to question previously: *Which values? Whose values?*

As a first step, it is worthwhile to justify this chapter's title. If one examines the *Oxford English Dictionary* for definitions of the verb "design" (as in, "to design"), the relevant entries make use of the following words: plan, purpose, intend, conceive, contrive, contemplate, devise, sketch, delineate, draw, mean, form, fashion. Essentially, "to design" is "to plan," "to purpose," "to intend," and so on. These words do well to capture our conventional, colloquial understanding of what the practice of "designing" entails. They convey the many facets of designing an object such as a motorcycle. A person (or a group of people) plans out the various components that determine what the motorcycle will look like, what it will be capable of doing, as well as a way of making of plans a physical reality (the motorcycle itself). Mechanical engineers will work to contrive such things as the structure of the machine's engine, its transmission, its control mechanisms, its lights, sensors, and indicators. Other designers will focus on the aesthetic components of the motorcycle: the shape of its fuel tank, its fenders, the arrangement of handlebars, the placement of lights and blinkers, the color scheme, the location of the maker's insignia. Some components will require cross-disciplinary design, such as the placement of spokes on the motorcycle's wheels—a decision that concerns both structural issues and visual appeal. Engineers concerned with manufacturing and assembly will plan the methods by which components are to be made and brought together into a final product. Many people will design, and do so in different ways, but their diverse practices will all fall under the group of things listed earlier. These words begin to uncover what "design" is, which in turn helps clarify what kind of person "a designer" is, as well as what "a design" is.

What makes the words found in the definition similar? What do these words imply that bears on what "designing" involves? Most importantly, each of these words is an *intentional human action*—that is to say, something that people do purposefully. Design is not a spontaneous happening. Motorcycles do not appear out of nothingness (they are constructed), and their final form is not a random occurrence (it is the result of planning). When a corporation like Harley-Davidson chooses to introduce a new model, a decision has been made to begin designing something. A process of planning, of devising, of thinking, of reworking and amending is carried out. Old and existing models are examined, ideas are explored, and requirements are specified. When those tasked with devising the new machine—individuals such as engineers and automotive designers—are satisfied, the process of designing comes to a close, and the related but distinct phase of manufacturing can begin. The choice to begin designing and every decision along the way occur for some reason. Design doesn't happen for its own sake: it is always motivated by something. A motorcycle is designed and manufactured for many reasons. Some of these reasons are pragmatic—people have become accustomed to using machines to travel across distances. Other reasons are economic (Harley-Davidson wants to continue selling motorcycles to consumers) or professional (motorcycle engineers may benefit personally from particularly well-regarded designs). The point is this: "designing" occurs for specific human reasons and is driven by human volition.

That design occurs for specific reasons implies that there are particular motivations for any given design project. A decision has been made to expend time and energy in what may be a long endeavor because the effort is deemed worthwhile. Whatever is to be made is considered valuable enough—worthwhile enough—to justify the effort. "Value" may encompass just the economic benefit of producing something (the profit to Harley-Davidson of selling motorcycles), but it also includes the less tangible (the passion of engineers for producing beautiful machines). Motivation to undertake a design process is the first way in which values underlie decisions to design and make things.

Values also play key roles in deciding how to design and make those things. Harley-Davidson occupies a unique place in motorcycle design because its aesthetic tradition is relatively well defined. Thus, even new models must sit within an existing tradition. Choices about form and performance are strongly shaped by the brand's history of artifacts. These are decisions that rest on particular values. Some may be very straightforward: the company's insignia must be plainly visible. Others may be more involved: a particular form for the handlebars is more aesthetically pleasing than another and fits in better with existing designs. Ultimately, the end product will look a certain way and perform a certain way. These are not arbitrary results. They are the outcomes of a decision-making process, most steps of which involve making value-based choices about varying options. Consider this curious example. The sound of a Harley-Davidson's engine is unique—it has become deeply associated with the brand[1] and the style of motorcycles it produces.[2] The company has even attempted to trademark the sound.[3] The sound is a result of the placement of the cylinders within the engine and the shape of the exhaust—mechanical engineering decisions with aesthetic consequences. The placement of mechanical components is a balance between performance, efficiency, *and acoustics*—the design persists because the sound of a Harley-Davidson's engine carries with it important connotations of power, uniqueness,[4] and thrill.[5] These values in part guide this facet of the machine's design. So important are these values to motorcycle consumers in general that specialized companies offer noise, vibration, and harshness (NVH) testing and acoustic design (figure 5.1).

The example of a motorcycle highlights two important features of design. First, design is purposeful. Designers actively and intentionally carry out their work. Second, that work is motivated and shaped by values. There are choices to be made, and those decisions involve the application of values to given problems.

Does Nature Care?

With this understanding of design, I can now ask: Does design exist in nature? Surely, the answer must be no. To suggest that nature designs, or that design exists in nature, is to suggest that nature purposefully makes its things, and that those things are formed because of particular values. That is, it is to suggest that nature is actively aware of designing and cares about what and how it designs.

For nature to be aware of anything, it would have to be some sort of agent; that is, an individual or entity with awareness and the capacity to act, to do things. But what kind of agent would this be? Nature could be the collection of everything, everywhere that is independent of human beings—the sum total of things for which human beings can take no responsibility for creating. Because this book explores biological topics, it might be useful to focus only on the totality of all existent living things; that is, biological nature. In either circumstance, these conceptions are of a *collection of things*, and not of an agent. Any attempt to discern an overseeing entity driving the "design" of natural things is mystical, theological, or simply a metaphor or personification of natural events. Mysticism and theology get around the issue by way of the supernatural, so they lend little help in this investigation, and the use of metaphor or personification implies that reality is different. In this case, it suggests that while it is possible to talk about nature "designing," that talk is only a useful way to think about a complicated topic, and not because this is the way things are. Without an agent, there is no awareness or intention. Without awareness and intention, there is no purposeful activity, and design is a purposeful activity.

Next is the issue of values. As a human activity, design is motivated by particular needs or desires deemed worthwhile. Many decisions during the design process result from the set of values—norms, mores, standards, principles, and so on—belonging to those responsible for the project and to those who may eventually use the object being built. The structure and working of the finished product will depend on value-driven decisions. Can the same be said about the things of nature?

The notion that nature "designs" most often concerns the process of evolution. Thus, talk of "nature's designs" most often refers to the results of natural selection. To say that natural selection "designs" is to say that the form and behavior of living things are a result of evolution arriving at an impressive solution to problems found in the environment. The example given by Alistair Elfick and Drew Endy in chapter 1 fits in with this perspective. A single seed found in a pinecone can reproduce a fir tree, which will have its own pinecones. I share with Alistair and Drew a fascination with the marvelous complexity of this phenomenon—with the awesome results of millennia of evolution. "Design" might work well as a shorthand to the much more intricate processes involved in evolution, but again, its use is metaphorical—convenient, but not accurate. After all, evolution is a series of natural occurrences, not an agent; without agents,

there is no design. There was no planner that set down the structure and behavior of fir trees or their pinecones. In many ways, this is precisely why living things fascinate us so.

Consider another example, the case of mangroves in tropical jungles (figure 5.2). Riding on a kayak or canoe through the waters of a mangrove—as I have done often in my home, Costa Rica—reveals the overwhelming complexity of the plant life occupying the space. Trunks, roots, and vines twist into and around each other, seeking to secure the resources needed for growth. There is a strange beauty in the maze-like entanglements that make up mangroves, but there is certainly no systematic order. No blueprints were laid down to map out the space or to specify the structure of each root. No planner drafted a design. And yet, the system works; plants live and reproduce. Nature has no design, but neither does it need designs or designers. This, again, is one way in which it inspires fascination.

There are other differences as well, of course. Evolution is blind; that is, the process of natural selection is not aware of where it will end up. The same cannot be said of design, which always starts from an awareness of what the desired outcome is. Even if the end result is very different from the designer's original intentions, there is nonetheless a goal to satisfy. Engineers and designers of a new motorcycle know that their goal is a motorcycle; they may specify further the characteristics of that artifact. In either case, the end is set down and known. Evolution is also blind to alternatives; it cannot survey solutions found in other organisms or in previous organisms, compare these, and choose useful ones as it sees fit. Designers do this all the time. Motorcycle designers can review existing models, view parts catalogs, look at color palettes, and choose or modify what they see. Most importantly, where human design is motivated and constrained by values, evolution in nature is entirely value-free. Evolution works to encourage structures and behaviors that increase the likelihood that an organism will survive and reproduce. Design is based on satisfying human desires and needs, which are in turn based on human-centered values, interests, and incentives. Evolution has no concern for the success or failure of organisms or species, but designers have a clear interest in the success of their products. Stated clearly: designers care, nature does not.

Synthetic Biology and Design

Up to this point, my argument can be summarized as making two points about nature and design. First, that there is no design in nature. Second, that biologists use metaphors as shorthand for elaborate systems of explanation. Together, these observations suggest a number of things about synthetic biology and its relationship to design and living nature.

Figure 5.2
Mangrove roots and branches. The interwoven, messy character of the roots may not conform to systematic, rational design, but it works. It is also strikingly beautiful.

Nature is not designed, whereas human-built things are designed.[6] Outwardly, it then appears that synthetic biology—a field that wants to build biological technologies systematically—hopes to make living nature "designable." A reading of Alistair and Drew's chapter earlier in this book would support this impression. The two synthetic biologists speak with exuberance about their goal of rigorously engineered biological technologies. As I see it, the observation that synthetic biology wants "designable" biology is somewhat problematic. Without any further claims or qualifications, it is somewhat unsatisfying and continues to create difficulties.

Consider first its superficiality. To say that synthetic biology brings "design" to living nature does not specify what precisely the field understands by "design," why it is motivated to undertake this project in the first place, what exactly it hopes to achieve, or how it makes sense of good and bad design. Each of these issues is hugely important. Without thinking through them, there is little hope of understanding what synthetic biology is, much less of evaluating its place within the broader scope of human activities. If we hope to say something intelligent about synthetic biology and critically judge its worth, then the statement "synthetic biology hopes to design with nature" should be little more than a starting point. Our discussion should focus on what precisely "design" presumes, involves, and brings about in this field.

Second, we must be mindful of the dangers involved in too loudly proclaiming synthetic biology's aim of designing with nature. Most importantly, doing so presents the danger of simply repeating synthetic biology's own rhetoric. After all, the field asserts that its novelty, its worth, its very justification, rest on the capacity to design with living systems. Unlike genetic engineering, which is dismissed as "mere" craft, synthetic biology is presented as a field that will design systematically—it will be "true" engineering. Even the most discerning analysis by outsiders runs the risk of exaggerating the substance of such claims. It can make of rhetoric reality, and unintentionally bring weight to promises. Observers like me generally note that synthetic biology aims to, or hopes to, or is working to design with nature. However, it is all too easy for such claims to be taken as statements that synthetic biology actually does design with nature. This is not the case. With time, the field may develop the necessary tools and techniques for systematic design with/of living systems, but it has yet to do so.

To avoid the first problem—superficiality—we must look past the bare notion of "design." To avoid the second—unintended meaning and advocacy—we must choose issues that allow for critical study of synthetic biology. Values represent one such topic. To state, "there is no design in nature," and to suggest that synthetic biology hopes to make of living things

a substrate for design, is in part to claim that synthetic biology introduces the question of values into a domain previously independent of them.

Synthetic Biology and Values

Synthetic biologists often portray their efforts as aimed at moving from "what is" to "what could be."[7] That is, a shift from scientific studies aimed at cataloguing and understanding what exists in the natural world to engineering work that makes new things possible from the materials of living stuff. Insofar as design does not exist in nature, but it does in human artifice, this shift is a valid portrayal of synthetic biology's goals. However, as I noted, it can only serve as a starting point for a critical discussion of the field. A more useful perspective on synthetic biology's ambitions looks to what desires such design hopes to fulfill.

Nature does not care about its things. Organisms evolve, survive, reproduce, and do all of the things associated with life, but not once are they

Figure 5.3
An image from *Cat Fancy Club,* a Somewhere project by Nina Pope and Karen Guthrie. Our domestic animals have been shaped by a range of human-centered values, including aesthetic ones.

subjected to evaluation, nor is a single trait ever introduced or changed because of the desires of some overseeing agent. Species come to be, exist and change, and some disappear. None of those occurrences are motivated or shaped by values. This is the very core of what is meant by "there is no design in nature." The development of nature's living diversity is a result of many value-free events. Thus, to say that synthetic biology wants to design with nature is to say that if successful, it will make the development of some living things a matter for value-driven decisions, rather than the independent workings of the natural world. The change synthetic biology hopes to bring about is not simply to add human-driven novelty to the existing diversity of living things. It is not simply a shift from "what is" to "what we make" (or "what could be"). It is one from "what is" to "what satisfies our needs, desires, and expectations."

Humans have carried out similar transformations for much of our history on this planet. The selective breeding of plants for taste and quantity and of animals for speed or strength or beauty are comparable instances of people bringing human-centered values to bear on the form and behavior of living nature. Nina Pope's *Cat Fancy Club* documents examples of domestic animals bred following aesthetics standards[8] (figure 5.3). What a cat show demands of animals is certainly not what a wild environment demands of feline species, and the results are living creatures shaped by human taste, rather than the challenge of survival and reproduction. Humans alive today are not the only ones who have sought to shape the world to their satisfaction. However, people today may seek to question the "why" and "how" of these changes. As I see it, synthetic biology's design aspirations present us with an important challenge.

The Challenge

Values are specific, as are their origins, applications, and ramifications. They belong to and reflect particular communities of people. As such, to say that design introduces values to a realm of things that is otherwise value-free is to imply that those things must satisfy the needs, desires, and expectations of a particular community. Our example of a motorcycle works well to illustrate this point. A company like Harley-Davidson works to satisfy an existing base of users, whose perspective is shaped by particular tropes of freedom, virility, and uniqueness. The form and behavior of the machine—its "design"—will reflect these values and concepts. Other enterprises in design feature the same role for values. A 1936 exhibition at the Museum of Modern Art (MoMA), entitled "Edward Steichen's Delphiniums," featured flower arrangements by a renowned artist and horticulturalist.[9] In this first living matter exhibition to be held at MoMA, Edward Steichen brought to bear artistic sensibility on biological nature; the values guiding his shaping of delphiniums were aesthetic and cultural in character. Broadly speaking, his project mirrors much of what synthetic biology seeks to make possible—human design of living nature based on human values and desires. The following questions suggest themselves: *Which values are to guide the design of living things? Should they? Whose values are they? What is the goal in trying to design with a biological substrate? Why are we pursuing this goal?*

As I see it, these questions challenge us to think seriously about what roles synthetic biology may play in our societies, what place or places humans should occupy in relation to the rest of the living world, and what underlies decisions about these roles and places. Since the advent of tools and techniques for intervening in the domains of genomes, there have been concerns and question about how we should use and direct our biotechnological activities. A great deal of that work has been assigned to bioethicists, whose contributions have been important but often abstract. I see the matter of ethics as far more grounded in the everyday of biotechnology. The questions should take less the form of "Should we play God?" and more the shape of "What intentions and assumptions are driving this design practice?" This type of question is less obtuse, more encompassing, and gets at humans' relationships to other living things without sacrificing the matter of ethics.

As matters currently stand, many synthetic biologists speak of "making biology easier to engineer."[10] Alistair and Drew's presentation of the field earlier in this book follows roughly along those lines: The two define synthetic biology as engaged in a very specific form of design—engineering design. Efforts at standardizing biological phenomena,[11] making biological parts,[12] and devising tools for systematic planning and fabrication of biological technologies[13] all follow norms, mores, and principles derived from existing engineering disciplines. These include the all-importance of usefulness; the desire for certainty and control; and the need for strict organization and concrete standards. Well-designed things are useful, reliable, controllable, and follow exacting parameters. While such priorities may serve traditional engineering well, there is no obvious, indisputable reason why they should guide the making of synthetic biological technologies. These values, as well as their predominance, require justification.

As different authors describe in this book, design in synthetic biology is by no means a set and fixed way of doing things. Indeed, one of our motivations was the uncertainty of what design means, entails, and looks like in synthetic biology. At present, many in the field hope to reproduce the design practices of conventional engineering.[14] This commitment does not stand on a thoughtful evaluation of engineering design, but rather on a reluctance to move away from tradition and a desire to be like other engineers. Crucially, it rests on an important claim: making technologies with living things is fundamentally no different than making technologies with inanimate materials. Using a cell to build is the same as using rocks and metals. This is not an essential, impartial truth of the world. It is a judgment of what is thought best and a choice based on specific values. Like any choice, it represents one course of action, but not the sole one. Engineering may serve us here as it has elsewhere, but this is by no means a foregone conclusion; it is a possibility.

There is no design in nature. Design is something that people may levy on the biological world, and synthetic biology is one particular attempt to accomplish precisely this. As with any design effort, synthetic biology is motivated by, draws upon, and is shaped by specific values. At the present, those values stem from traditional engineering disciplines. As synthetic biologists and outside observers such as those featured in this book explore what "designing" will entail for this emerging field, a necessary point of discussion must be this: Which values will inform if, why, and how we make biological technologies?

6.
Design Evolution

Alexandra Daisy Ginsberg

In his lab in the Department of Plant Sciences at the University of Cambridge, its benches strewn with Petri dishes full of bacteria more typical of a microbiology laboratory, biologist Jim Haseloff was showing me old BBC footage of pitcher plants growing in slow motion.[1] The pitcher plant builds itself a complex form from a patch of dirt, a liquid-filled vessel to trap its prey. Jim pointed out the little fleshy nodule appearing at the end of the leaf, snaking toward the ground, finding a level surface and doing a U-turn, then inflating in all directions to build its pitcher, complete with lid, like in figure 6.1. With time sped up, the unique material properties of biology exposed are somehow surprising. We often forget biology's ability

Frontispiece
E. chromi: The Scatalog

to convert sunlight into chemical energy, build complex structures from scratch, its adaptive, self-replicating, self-repairing capabilities, and use of its own waste material to fuel new structures, eclipsing any notion of human ingenuity.

During this first conversation with Jim back in 2009, as he explained his work in engineering plant cells, I began to see the potential that biology has to challenge our ways of thinking and of consuming, conceptually if not literally. Jim is a synthetic biologist and researches plant morphogenesis: how plants grow as their cells differentiate. But his unspoken aspirations appeared to be bigger than improving vitamin production or hardiness, the travails of classic plant genetic engineering.

Figure 6.1
A fanged pitcher plant, or *Nepenthes bicalcarat*. The pitcher plant grows its own vessel, complete with lid, to trap food.

At the Royal College of Art in London, over the preceding weeks, fellow artist Sascha Pohflepp and I had been dreaming up disruptive manufacturing systems that use biology. This conversation had begun over science fiction classic *Soylent Green*.[2] We were curious as to how system collapse can force a technology into action (though *Soylent Green*'s solution is a little extreme, the eponymous rations feeding an exploding population turn out to be made from those same people). Could a collapse ever trigger exceptional progress in synthetic biology, for example if oil shortages made our current system of global manufacturing and shipping of goods impossible? Pursuing this idea, we came across the *BioFabricator*, Imperial College London's 2008 iGEM attempt to engineer bacteria into a light-activated three-dimensional printer, which eventually led me to Cambridge to visit Jim's lab.[3]

I asked Jim where he thought synthetic biology with plants might lead. At this point, I was under the impression that microbes were the field's favored organisms. He paused, then replied, "One day, we might grow products inside plants." It was deliciously ambiguous. Did he mean chemical products or was he really hinting at controlling form, like the growing pitcher plant, frozen on his computer screen?

Back in the studio, Sascha and I followed this line of thinking. Would it be possible to program and control plants to solve today's unsustainable consumption patterns, using their natural logic of self-organization and modularity? Would living organisms become industrial robots, replacing mechanical production lines? Exploring the ramifications of such massive system change, we designed a commodity from this potential future to understand it better, using design as a means for investigation—not prediction—of future products.

This became *Growth Assembly*, a product that would be grown in seven different parts, its components harvested and then assembled.[4] We decided that our prototype consumable would be an "herbicide sprayer," an essential device in this future. Everyone would need one to protect his or her crop of delicate, engineered horticultural machines from the threat of existing nature, which through the processes of evolution can naturally defend itself.

With a blueprint drawn—a cross-section of the final product—we worked with illustrator Sion Ap Tomos to flesh out our plant designs, using the visual language of botanical illustration and inspired by the drawings of naturalist Ernst Haeckel.[5] When I e-mailed Jim for advice, he gamely reminded us to consider what he saw as "the constraints inherent in a scheme for manufacture by biological development: the processes of growth constrain topology, shape and scale; and that cell differentiation produces the material structure." We tried to incorporate these concepts as we designed the form of our imagined plants.

The seven grown components of our assembled product include the "herbicide gourd." As big as a melon, its juice is the herbicide (figure 6.2).

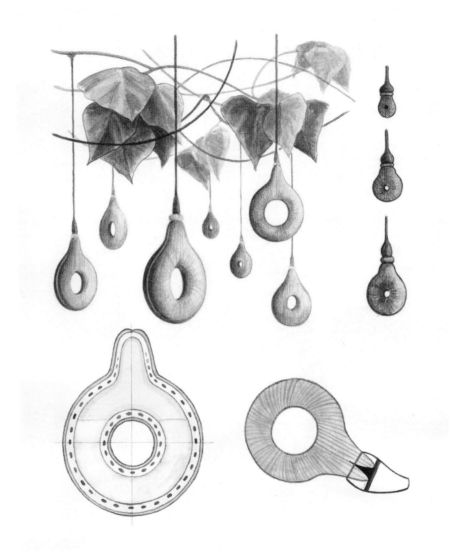

Figure 6.2
Herbicide Gourd from *Growth Assembly* by Alexandra Daisy Ginsberg and Sascha Pohflepp (2009).

Figure 6.3
Nozzle Fruits from *Growth Assembly* by Alexandra Daisy Ginsberg and Sascha Pohflepp (2009).

The thorn of the "spike fruit" is used to pierce the gourd's base; the user puts his or her arm through the hole in the fruit to carry the sprayer and squeeze the herbicide poison down through the device. The "connector" attaches the "spike fruit" to the translucent "tube," it grows inside a shell, like a walnut, while the tubes are cultivated as long, hollow roots. "Handles" are grown upside down, taking advantage of gravity to extend their length. "Nozzles" are picked from vine fruits related to the tomato (figure 6.3).

Grown, harvested, and assembled, diversity and softness are introduced into a realm that today is dominated by hardness and manufacturing standards (figure 6.4). Shops would evolve into factory farms with licensed

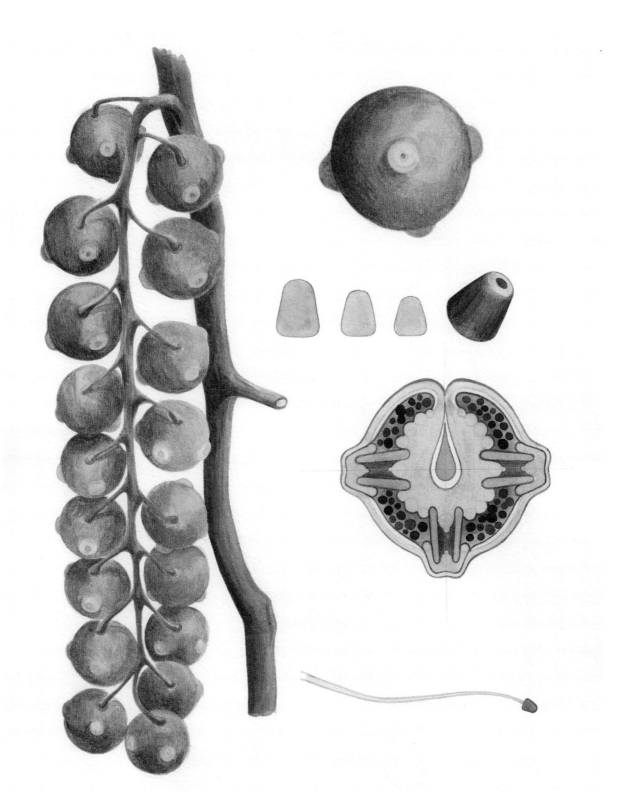

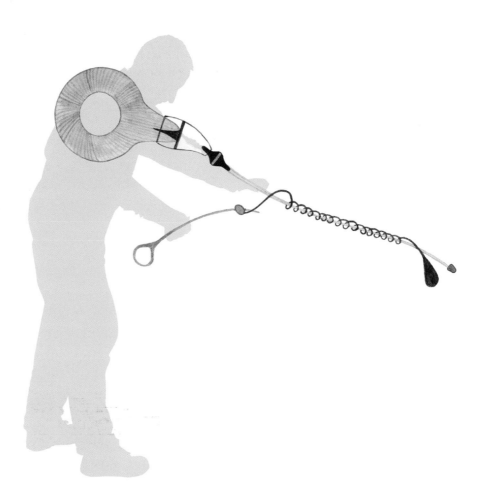

products grown and sold in the same place. We would no longer need to ship products, only seeds, as all the manufacturing instructions would be encoded into their DNA.

It is this aspect of the fictional *Growth Assembly* that seems so enticing. DNA could even be sent digitally instead of physically, printed and inserted into a system elsewhere, facilitating new modes of manufacture and distributed production to reduce waste and energy consumption. Our investigation inspired members of Jim's lab, including Fernan Federici, later a Synthetic Aesthetics resident (see chapter 7), and marked the beginning of my own collaborations with synthetic biologists, as well as Sascha's (see chapter 15). Yet there is a flaw designed deep in *Growth Assembly*. Industry subsumes nature entirely as human design strategies replace nature's own. Our seven designed plants, assembled into an "herbicide sprayer" to eliminate natural species, clumsily do what evolution elegantly streamlines: self-defense from nature's threats.

Figure 6.4
Assembled Herbicide Sprayer
from *Growth Assembly* by
Alexandra Daisy Ginsberg and
Sascha Pohflepp (2009).

While synthetic biology presents biology as an ideal future technology for making things, its unfolding industrialization reveals the pursuit of a different, human logic: biology reduced to bags of enzymes pumping out oil, locked in stainless steel fermenters. Unfamiliar crops like those imagined in *Growth Assembly* need not materialize, with any transition to manufacturing materials using biology imperceptible for the average consumer. The only evidence would be the new feedstock infrastructure grown in faraway places, supplying mountains of sugar shipped in to fuel vats of microbes. Considering public abhorrence of genetic modification (in Europe, especially), what could be better than a promising, green-tinted technology, with the living stuff safely locked away? Biology's unpredictability combined with societal attitudes, regulatory control, and safety concerns means that letting synthetic biology's products loose for widespread environmental release may well be a bad idea. Yet *Growth Assembly* points at uncertainties that locked-down microbes also present; for example, should we be using plants to make things and fuel, rather than as food for us? Within the current framework, the feedstock supply chain is not considered part of the biological design. And ultimately, *Growth Assembly* challenges us to ask whether our existing understanding of design and consumption can truly be applied unaltered to biology.

While the matter of what makes "good" biological design remains unresolved, synthetic biology's champions press on with their dreams. The White House launched its *National Bioeconomy Blueprint* in 2012, outlining a desire to shift from an industrial to bio-industrial model.[6] The idea that by using biology, a nation could transcend its existing system of exploitation and disposal, replacing it with a sustainable one, is noble. But this plan fails to define what sustainability or best practice actually means in this context. With a potent combination of political enthusiasm and the exponential growth of synthesis and sequencing technologies driving biological innovation ever faster, it is now that we must consider what we want and how we might change our ways, if at all. Thinking in more detail about what it means to design biology *well* could help to avoid repeating past mistakes, like the unintentional spread of genetically modified variants and their knock-on effects, both social and environmental. A biotech future may be far from *Growth Assembly*, but developing a deeper understanding of what design in biology entails can help us understand with greater clarity both what is possible and what is at stake.

Principles of design do exist in synthetic biology: not least, abstraction, decoupling, and standardization, the foundations of the approach.[7] Yet these are engineering design principles used to structure material and ideas. Designed objects do not live in a "perfect," idealized world: buildings move from the context-free three-dimensional environment of computer-aided design (CAD) programs to be lived in, used, adapted, but above all, experienced. Synthetic biology's story is of simplicity, but the complexity of life cannot be ignored. Since biological design cannot be isolated from the larger ecological and economic systems it is part of, how do we want biological products to operate in the world? These kinds of design principles are not rules like "abstraction," but considerations of how we might (want to) experience synthetic biology.

Up until now, design has been in pursuit of production, generally indifferent to the environment, as discussed in chapter 3. As we prepare to design evolution, our existing assumptions are challenged: We cannot avoid evolving design. Thinking more inclusively and extensively may offer a better way to design for societies, not just service parts of them. Instead of forcing design on the raw material of life through the design *of* biology, can synthetic biology design *with* biology? This would mean incorporating some of its most remarkable "design" properties, like life, death, reproduction, evolution, and self-adaption, rather than trying to design them out. This is not a rallying cry for biomimicry, designing in imitation of biology's form, but a call to design with the logic of biology. In chapter 7, architect David Benjamin and biologist Fernan Federici highlight this pursuit of "bio logic" at the cellular level. Biology stores information at many different scales, not just in the molecules of DNA. It functions from the microscopic to the global: from protein manufacturing to population control. At each intermediate scale, we must consider the interactions between biological objects and contexts and the people or systems or living things we are designing for. All these are integral to a new biological design paradigm.

Design and art can offer useful new insights as we develop these kinds of design "principles." The discussion and stories that follow in this section are intended to provide tools and insights into what "good" design might be. This is not a definitive list, but the beginnings of a debate about what a well-designed synthetic biology might entail, beyond merely technical considerations. These include what new archetypes it might produce, and what new ways we might conceive the life cycles and ownership of products and the relationship between them, the world, and ourselves. Building on the work of scientists, artists, and designers, it also shows how valuable these broader perspectives can be for the design of a technology promising to meet future human and environmental needs.

Good Design of New Archetypes

Design is concerned with the everyday functioning of the world around us. We are all design experts, developing our critical skills through our daily interactions with design. Consider two underground system maps or two jugs. One map may be easier to understand but might still get you lost; one jug seems more beautiful but still spills every time you pour. Which one is "better"?

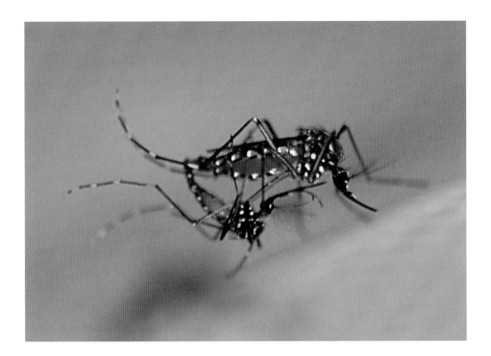

To judge these things, we intuitively evaluate utility, craftsmanship, cost, and more subjective metrics, developing personal parameters of aesthetics or style. In meeting our individual needs, good design can even feel invisible, marking the successful translation of a good idea into a well-designed object.

In their search for optimal designs, some contemporary architects simulate biological evolutionary processes, using "genetic algorithms" to design. These programs churn out possible architectural solutions within designed parameters. The generated designs lead to a "utopia point," an imaginary place where the perfect design lies, perhaps maximizing local environmental conditions or giving the most efficient form. Yet while there is good design, there is no perfect design. Design reflects a balance of constraints that includes changing needs and desires over time: The jug that spills can still look good. The concept of the design utopia denies the reality of individual variation; it assumes that we are all the same, all the time, and that there is a single best solution to a problem. Good design is personal and has elements of the poetic; such emotive or aesthetic parameters are hard to program into a computer algorithm. The eventual "compromise solution" is ideally as near as possible to the unreachable "utopia point," but it will be found through choice. The architect's selection from a generated set may well draw on personal, instinctive preferences, usurping the place of digital logic in the system.

Why does this matter for synthetic biology? Existing object archetypes—the jug or the map or the building—are designed products that we learn to

Figure 6.5
Oxitec Ltd RIDL male mosquito mating with a wild female. The male mosquito is designed so that his progeny will never be born, with the aim to reduce the spread of dengue fever.

evaluate from our previous interactions with them; we know what we do and don't like, what works for us, what doesn't. But designed biology proposes new functions, which are as yet hard to judge. We do not know what the technology might look like, feel like, or how it could affect our everyday lives, making it difficult to decide what is a good idea or not. Up until now, we haven't thought much about the design of the pharmaceuticals we use or the plastics we consume. But we may want to consider designed biology that we ingest or introduce into the ecosystem as design objects subject to personal choice, not just as rationally engineered machines of assumed universal benefit. What criteria can we use to evaluate these living products in real-world environments? We need to identify emerging product types before we can define the parameters and principles to learn to understand them.

Life and death are perhaps two novel design typologies of synthetic biology. Designing the stuff of life and death is not new in itself: obstetric forceps and pistols facilitate both. But in synthetic biology, life and death are "applications" designed in DNA, the biological material also the enabling "hardware". The mosquito is becoming one such product: British-designed RIDL male mosquitoes whose progeny are designed never to live. Engineered by Oxitec Ltd., RIDL males are being touted as a tool in the global fight against dengue fever, a mosquito-transmitted virus that infects more than 50 million people each year in tropical and subtropical regions.[8] Already on trial in the Cayman Islands and Brazil, factory-grown RIDL mosquitoes are sorted by sex, and then released into the wild by the million. The engineered males mate with wild females, whose offspring will never hatch (figure 6.5). Replacing previous eradication methods such as large-scale irradiation that renders mosquitoes sterile or chemical spraying that kills them, it appears to be a polite, British design solution to the tricky business of eliminating pathogens on a global scale. Mosquitoes aren't killed as such; they are just never born. Because the carriers of dengue fever are just one of many mosquito species, the RIDL designers argue that their extermination won't affect the ecosystem. Mosquitoes that never live seem to be good design.

In another national design stereotype, Swiss clock engineering is becoming biological. The Fussenegger Group at ETH Zurich is designing precision-timed bull sperm.[9] The sperm are encapsulated in an engineered cellulose sac before insertion into the cow's uterus, where they can survive for up to three days, far longer than normal. When the cow ovulates, her hormones trigger the capsule's disintegration, permitting conception (figure 6.6). This sperm-delivery system, designed in response to Switzerland's stringent regulations against genetic engineering, is already being tested in Swiss cows. Such a product coming to market could allow farmers to reduce the widespread use of hormone treatments used to synchronize herds' ovulation, saving time and

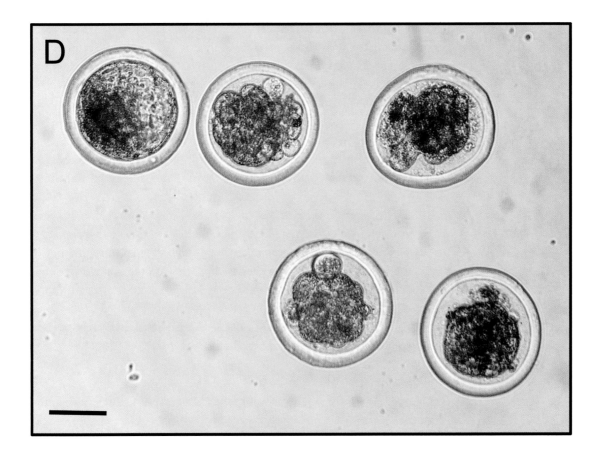

cost in monitoring the fertility cycle of individual animals before expensive insemination. For farmers at least, the engineering of reproduction appears to be good design.

These are some of the first nonmicrobial experimental products from a field that has so far been dominated by bacterial-scale endeavors. For the consumer, the only experience of them may be far from the site of intervention and intangible: A reduced chance of infection or cheaper cheese. But when the complexity of a design's interactions with the world is beyond computation, how do we evaluate its quality? Eradicating mosquitoes may be desirable for us in the short-term, but is it good longer-term? Past experiments at pest control have on occasion turned out catastrophically. Myxomatosis introduced to temper rabbit populations in France in the 1950s has been the possible cause of the near extinction of two of its predator species, the imperial eagle and Iberian lynx.[10]

Beyond field risk analysis, these design archetypes pose other questions, some unprecedented. "Design" transfers beyond the original object: Consumption is neither immediate nor a matter of choice. A biological product

Figure 6.6
Fertilized eggs removed from a cow's uterus after insemination using engineered encapsulated bull sperm. The Fussenegger Group at ETH Zurich is finding ways to control the release of sperm.

whose function is life or death may at first take be identical to the non-engineered organism. A new feature may be a function of a tiny intervention in a sequence of nucleotides. The engineers of RIDL mosquitoes design them to be indistinguishable from unmodified rival mates. This changed genetic code is not ephemeral like digital code: pull the plug out of the socket and it is gone. The organism becomes not only a living thing that looks like any other, but also one that can impact the life and death of others over generations or even across species and ecosystems. And these early experiments in insects and cows may also presage synthetic biology moving into other complex organisms, like us.

While design objects that cause life or death are not new, biological ones present questions we may not have considered before. As we invent new product archetypes and seek parameters for good biological design, we must also define the boundaries of ethical design. If the "utopia point" is an unattainable design ideal, we should be aware that for biology, the "compromise solution" may vary wildly, depending on your vantage point. We must question what we make, but particularly for unfamiliar designs that might be invisible, even intangible, but have far-reaching effects.

Leases of Life and Death Dating

Products die. Not literally—yet—but when they come to the end of their functional life, they are effectively dead to us. We discard our worn-out trainers and out-of-date phones, their replacements perpetuating our economic system. Here lies an inherent contradiction: capitalism's inclination toward unbounded growth, fueled by a finite planet. The balance required for sustainability has so far proved hard to achieve.

The origins of product death date back in part to 1932, when Bernard London, a Manhattan real estate broker, tried to address the problem of endemic economic depression, seeing years of technological progress and production being undermined by a lack of money to buy the surplus of stuff. In *Ending the Depression Through Planned Obsolescence*, he explained the issue as a matter of "organizing buyers" rather than consuming less.[11] "The new paradox of plenty constitutes a challenge to revolutionize our economic thinking," he wrote, arguing against the Malthusian view that humanity's increasing population is at risk from the planet's limited food production capacity.

At that time, the United States had replaced the excesses of the 1920s boom with the opposite extreme, a mentality of make-do-and-mend. To rebalance production and consumption, London suggested that objects should be sold with a legal document, a "lease of life." When their allotted time was up, they would be pronounced "legally dead" and "would be controlled by the duly appointed governmental agency and destroyed. . . . New products would constantly be pouring forth from the factories and marketplaces, to take the place of the obsolete."[12] Keeping products too long would incur financial penalty. Tax would be collected when an artifact "dies," not its owner. For London, both consumer products and humans had life spans, and the state should have the authority to define that span, for products at least.

The once radical argument that it is more cost-effective to destroy obsolete goods than risk human well-being underpins contemporary consumer desires. It assumes, incorrectly as we now know, an abundance of raw materials and an environment independent of human survival. Yet short-term economic gain still wins today. London's obsolescence never become state-enforced, instead, we allow corporations to instruct us when to dispose of things. Obsolescence is entrenched. "We joke that we design landfills," a designer quips.[13] Psychological obsolescence (clothes going out of fashion) and functional obsolescence (ditching minidiscs for CDs) are accepted. Meanwhile, secret obsolescence is a hotbed of conspiracy theories, beginning with the 1920s lightbulb manufacturers' cartel, accused of engineering their bulbs to fail sooner.

Value engineers usefully design your car's components to fail around the same time. There is little point making each part last as long as possible; the skill is in ensuring that the first part to fail does not kill the system and can be replaced. Less complex products are now designed with the price of repair a little below that of replacement—a broken laptop screen costs just less than a new laptop. Some products cannot be fixed: batteries are irretrievably encased within iPods. Product death today is a programmed inevitability. Even artificial implants in our bodies—hips, breasts, and chins—have to be replaced, their calculated life span reducing risk. Obsolescence accustomed us to products having designed life spans, some ever shortening. This entropy is normalized by the idea of technological progress.

While obsolescence in inanimate technologies may not have got as far as mandatory "leases of life" or "death dating," these may be significant concepts for the control of living machines. Living products could control the life or death of others; and we will have to control their lives too, designing products that die. Biology is already subject to a natural obsolescence of sorts through evolution, enabled by reproduction and death. Darwinian survival of the fittest is even referenced in the early literature on industrializing obsolescence. But evolution differs from design's deliberate strategies. Fish swim bladders that evolved from lungs—enabling fish to live symbiotically with worms—are happy accidents.[14]

By killing organisms before they reproduce too many times, synthetic biologists would hope to eliminate evolution and hence undesired biological obsolescence. Kill switches and terminator genes are currently seen as the best tools to control designed organisms. A kill switch is a suicide instruction written in DNA against malfunction or unwanted population growth. In theory, it could be applied to bacteria released into the ocean to clean up oil spills, which would self-terminate once their job is done.

Kill switches could also be a means to design functional obsolescence into biology. Such application may well anger consumers, as farmers worldwide have experienced as they encounter bans on saving GM seeds or compelling them to buy new hybrid seeds each year. In 1999, Monsanto pledged not to develop "terminator" seeds that yield seedless plants, but reserved the right to develop means to "inactivate only the specific gene(s) responsible for the value-added biotech trait," preventing it from being passed on to the next generation.[15]

This all may be moot: kill switches are as yet unpredictable in their effectiveness. Evolving, living designs still select for the fittest. Organisms not only swap genes; they will reject inserted genes that hinder them. Designing a bacterium to fail after 100 generations could reduce the risk of perpetuating faulty DNA, but it does not eliminate that risk. We may end up creating selective pressures for organisms that can avoid being killed by us.

Designs that are allowed to evolve may be the most unique way that functional obsolescence could play a role in synthetic biology, as described in "The Well-Oiled Machine," a short story I wrote with Oron Catts. In this future, evolution is used as a design tool, and constantly changing products generate unprecedented symbiotic marketing strategies.[16]

Obsolescence in products has proved damaging for the environment but good for technological progress. If living biotechnological products become prevalent, how will they slot into existing consumer systems? London's state-enforced death dating and leases of life may be useful references as we reconsider our ownership of and responsibility to living products. A state-mandated "lease of life," a prerequisite to kill biological products after a number of generations, could offer an economic and social safety mechanism beyond the kill switch, as we continue to seek reliable ways to design a good death for our things. While biotechnological obsolescence may become a fact of life, a functional design reality, it also marks the ultimate instrumentalization of life.

Life Cycles Not Life Spans

We tend to think of objects as isolated in their form and function from their environment, but objects do not operate independently. A computer interacts with other computers within networks, it sends and receives information, its invisible algorithms manifesting in physical evidence elsewhere: traffic lights change, online shopping arrives, stock markets lurch. The computer's form is not detached either. The machine consumes electricity from the grid, generates heat, and is built of materials each with their own history and net-work of impacts beyond the object on our desk.

We are beginning to reimagine products beyond their physical func-tion and form: flights are sold with optional carbon offsetting packages, nominally recognizing something less tangible as part of the product bought. Yet overwhelmingly, product life spans, not life cycles, continue to be the norm in product design, an issue raised by ecological design activist Wil-liam McDonough. In his manifesto *Cradle To Cradle*, he identifies the worst offenders as "monstrous hybrids": products, such as trainers, which combine "technical and organic nutrients."[17] Their component materials are so deeply integrated that they are impossible to separate for recycling. Landfill marks the inevitable end of their life.

Efforts to reduce consumption, or make new things from waste prod-ucts, or avoid problematic materials have had little impact on a global scale. China's National Stadium was built for the 2008 Beijing Olympics from a massive 42,000 metric tons of steel.[18] Manufacturing 1 metric ton of steel emits nearly 2 metric tons of carbon into the atmosphere. Yet alternative structural materials are available. A tree actively reduces atmospheric carbon levels during its production. "A ton of timber is in effect hoarding 1.4 tons of carbon," explains architect Alex de Rijke, who pioneers building with engi-neered timber. For him, "timber is the new concrete," or more specifically, structural timber made from layers of lumber glued together with nontoxic glues, which can be composted, and could even be used to build stadiums.[19] Biology's vast library of materials, derived from the same limited chemical palette, is easily decomposable for reassembly into new things. Engineered timber is a good model for life cycle design that fully integrates a product's existence before and beyond its functional life.

Yet synthetic biology appears to be making its own "monstrous hybrids" to meet demand, designing bacteria that pump out non-biodegradable acrylic acid for plastic or isoprene for tires. Once they leave the factory, these plastics and tires may be no less polluting than conventionally made ones. Could syn-thetic biologists instead invent novel self-contained production systems? These biological systems would integrate energy, material, manufacturing, assembly, and disposal; waste would be not only biodegradable but also useful for making the next generation of products. Since biology naturally makes consumption useful instead of ecologically harmful, synthetic biology could set a precedent in manufacturing by taking control of the entire design life cycle.

Artists, designers, and companies are innovating with biological materi-als in this way. Suzanne Lee, a British fashion designer, has pioneered the use of the fermented drink kombucha—a symbiotic mixture of yeast and bacteria—to make cellulose.[20] Lee manufactures biodegradable clothing

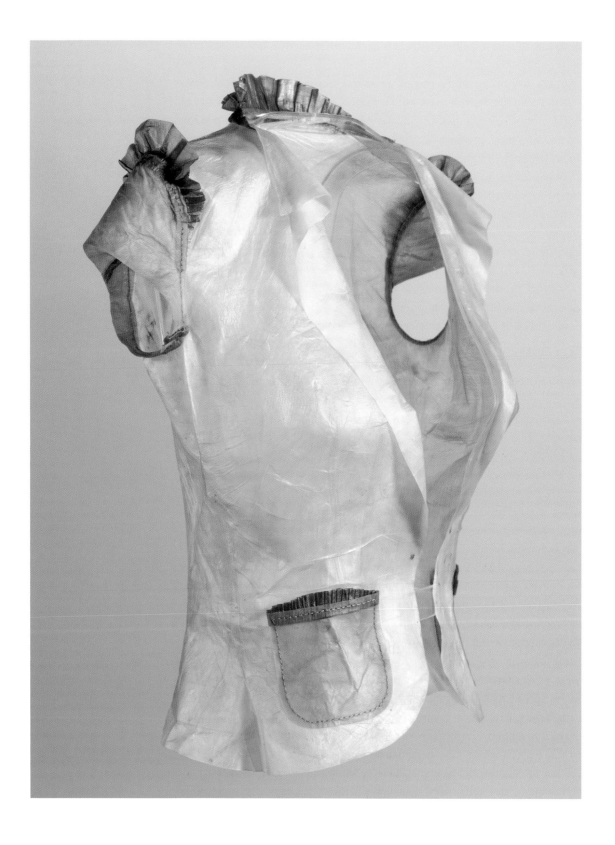

from the resulting flat, dried, leathery material and is seeking ways to improve the durability of the cellulose and innovate its production methods (figure 6.7). Through her *BioCouture* research, she has made the material accessible for other designers to experiment with. In *Living Artefacts* and *The Kernels of Chimaera* (both 2012), Stefan Schwabe attempted to coax new forms from the material, growing endless rolls, bubbles or connected layers, and automating their growth.[21]

Mushrooms are being experimented with as construction materials too. American artist Phil Ross builds *Mycotecture*, functional structures made from cast mushroom bricks (figure 6.8), while Ecovative Design is growing fully biodegradable and compostable packaging from mycelium for consumer products like televisions and wine bottles.[22]

While we may want designed things to stay the same, they inevitably fall apart, decaying over time. But entropy could be a useful design feature for biological products. Artist Jae Rhim Lee's *Infinity Burial Project* includes a

Figure 6.7
Suzanne Lee's *BioCouture* jacket, made from kombucha-manufactured bacterial cellulose, is an example of her ongoing experiments with this biodegradeable material.

Figure 6.8
A single *Mycotecture* brick of the *Ganoderma lucidium* fungus (2009). The U.S.-based artist Phil Ross is experimenting with using mushrooms as a building material.

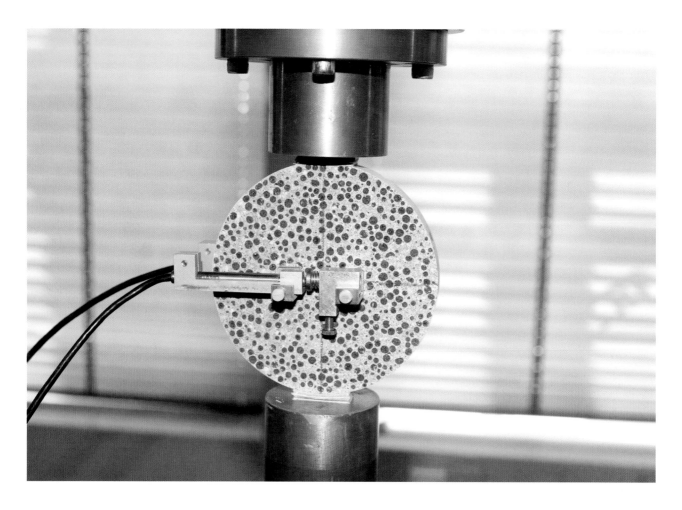

Figure 6.9
Dr. Henk Jonkers's *BioConcrete* in a testing rig. The self-healing concrete is impregnated with *Bacillus* spores that can lie dormant for up to 200 years, which could help increase the concrete's life span. As water penetration activates the bacteria, it triggers them to turn nutrients added in the concrete into limestone.

prototype *Mushroom Death Suit*, impregnated with mycelium spores.[23] Her intention is to design a variety of fungi that can decompose bodies effectively and also remediate soil of the toxins that we accumulate during our chemically infused lifestyles. The *Mushroom Death Suit* challenges accepted practices of preserving bodies for burial. Instead, it promotes decay, and highlights an often hidden impact of our own life cycle on the wider ecosystem.

If we can reimagine how products age and wear down, how they die and will be recycled, can we imagine how they could usefully change over time? Up until now, stasis in the design of objects has generally been expected: A car that repairs itself might be convenient, but a car that evolves to meet its own needs could be problematic. Designing with change in mind, investigated further in chapters 15 and 16, is a curious conceptual leap for the designer. Adopting this unfamiliar property, instead of trying to eliminate it, could yield interesting directions for design. Self-repairing materials are the first step, as microbiologist Henk Jonkers's investigations into bacteria-infused self-healing *BioConcrete* show (figure 6.9).[24] Could biological designs go further, usefully self-adapting from their original design? Such machines would certainly present novel risks.

Richard Lenski's *Long-Term Evolution Experiment* has been tracking 12 originally identical *Escherichia coli* populations at Michigan State University since 1988. In 2010, the project celebrated 50,000 generations. The meticulously documented saga reveals each population evolving along its own path, although their identical contexts ensured patterns of similarity.[25] Synthetic biologists designing with biology take advantage of this ability of organisms to iterate over time, generating thousands of possible solutions to a problem. Optimum designs are then selected through screening, just like the architect using genetic algorithms. But unlike the building, the design of the organism does not stop changing once it has been chosen. Biology, as Lenski's elaborate project demonstrates, continually evolves, adapting to context, self-optimizing over generations.

Context can be used as a design tool: Synthetic biologists can design the environment as well as the organism, effectively collaborating with biology to design. For the designer, relinquishing control to co-author with the material is unusual. But in biology, designing an environment that, for example, encourages very precise behavior can allow for that behavior to emerge, and persist. This method was used to design the Endy lab's bacterial data storage system to remain precise: after 100 generations, the selected design was unchanged.[26]

To be reliable, biological design may not always need to be precise. Reliability and precision are different things. A T-cell, a kind of white blood cell, could be engineered as a therapeutic drug with a counter in its code to limit its number of replications. Counting down from 100, it would then self-destruct. Some "noise"—random fluctuations—may be acceptable: If 100 divisions are allowed, there may be no significant difference if 99 or 101 divisions take place, or even 90 or 110. The fact that the system can be controlled and is reliable is enough, which contrasts with the precision of digital computation and challenges traditional engineering concepts. Noise need not be designed out; it can improve the function and adaptability of biological systems.

Since the invention of the assembly line, we have come to value the uniformity of products. Experimental materials, from the cellulose grown from a kombucha "mother," to concrete that fills its own cracks over time, suggest a new biological aesthetic, a visual and experiential language for biological products. They challenge existing systems and assumptions about mass production and may even provoke discomfort caused by an entrenched cultural disgust at biology out of place, a phenomenon examined in chapters 17 and 18. Synthetic biologists' ambitions of standardization may lie at the scale of DNA, but the tangible products and materials manufactured by biology, from plastics to cosmetics, are being designed to be indistinguishable from those produced by synthetic chemistry. Could future goods benefit from looking different to those made by production lines, celebrating the aesthetic diversity of biological materials?

Control over the life and death of products is part of a much-needed paradigm shift in our understanding of designed things, a shift that is not only compelled by the ambitions of synthetic biology. Product footprints extend past objects; the things we make cannot be isolated from their contexts. Replacing short-term attitudes—where products and their life spans are conceived of as separate from the environment—we need to start designing product life cycles, with change and decay in mind. But with designed product life cycles might come unexpected compromise. Whether self-adapting, biodegrading, or simply visually irregular, could we bear to change our consumer behaviors, interactions, and lifestyles to accommodate such inventions?

Feeding the Machines

There is another aspect of the design of biological product life cycles that cannot be ignored: biology needs to eat. Beyond low-tech mushrooms or kombucha, others are experimenting with bioengineering to grow "sustainable" materials. Modern Meadow, a research start-up from the co-founder of Organovo, a company with expertise in printing cells, is trying to grow, among other things, leather from tissue culture. This process has supply chain dilemmas; as described in chapter 2, tissue-cultured cells gorge on a serum made from fetal calves. Finding ways to use these techniques for mass production still needs to be resolved.

It is not just unusual or expensive feedstock that is problematic. Lee's kombucha lives off sugary green tea, while biofuel-producing microbes eat sugar alone. The bioeconomy is sold as a sustainable, glucose-powered future, a sweet medicine to remedy our dirty carbon and dangerous nuclear habits. Living microscopic factories are seen as a weapon against our reliance on oil in the fight to keep economies running. It is hoped that switching from ancient organisms stockpiled in crude oil to newly sequestered carbon in biomass will provide the gamut of synthetic plastics and rubbers and chemicals we rely on. Grown in tropical plantations in Brazil and Indonesia, the biomass is shipped to feed brewing vats filled with engineered organisms. We may abandon grubby energy for this green paradise, but a scaled-up "Glucose Economy" may not be as sustainable in the long-term as it sounds.

The swaths of land converted to sugarcane production are sometimes justified as otherwise useless.[27] This assumption affects the communities that live there, but it also seems flawed from a design and supply chain perspective. DNA-scale design decisions have global impact when the systems are

scaled-up: biomass needs shipping, on a large scale. Whether replacing virgin forest, losing biodiversity and increasing chemical and water pressures, or simply replacing food crops for fuel, we must ask: Is there enough land to feed planes and cars and products as well as people and animals, and can such large-scale monoculture farming be sustainable?

Energy conversion is difficult. The more steps added into this process, the less efficient it is. Photovoltaic solar panels struggle to recoup as much energy in their lifetime as is used in their production, compared to plants and organisms that are naturally highly effective at harvesting energy from the sun through photosynthesis (this is the origin of the energy in oil and coal). The scientists engineering algae to convert solar energy directly into fuel are hence attempting to reduce the number of steps between harvesting and conversion, compared to the biomass feedstock system. But scaling-up algal fuel production comes with its own challenges.

Imagine a vast agricultural plantation stretching as far as the eye can see (figure 6.10). Rolling gray-blue clouds set off the green, a verdant stripe in

Figure 6.10
Many algae farms already exist, safely farming nonengineered algae for food and fuel.

an otherwise drab landscape. Open ponds are filled with green algae; in the concrete-edged tanks, iridescent smears on the surface shift with the movement of the clouds. A few workers travel up and down, occasionally scooping up a sample of slime, placing it into a rack for testing.

The engineered algae on this coastal stretch, in some not-so-far-off possible future, are excreting the same liquids that fuel our cars, as oily slicks ready to be skimmed off. The algae takes in CO_2 and breathes out clean oxygen. They may even be optimized to fight infection, or variation in outdoor temperature, or designed to adjust to seasonal light levels, and self-adapt to local water chemistry to minimize operation costs.

While this setup seems laudably green, an improvement from our dependence on extracted oil and more sustainable than the sugarcane-as-feedstock conundrum, it poses other issues. Picture those bruised clouds above unleashing a monstrous storm. The green soup is sucked up and rained into neighboring waterways, over fields and farms. The algae, robust and ready, have been approved safe for outside use. They make a happy home in local waterways, already full of fertilizer run-off. The streams become emerald swamps, glossy with oil.

This fiction reveals the inherent, though not necessarily intractable, complexity in the bioeconomy dream. Biologist Patrick Boyle, engineering algae in the lab and realizing these design problems, initiated a pilot workshop in 2011 to bring together experts across disciplines and scales, from DNA to ecosystems, to inform his design thinking.[28] This two-day workshop was a form of design "crit." Bringing together specialists from fields such as ecology to critique a design in synthetic biology is, as yet, surprisingly unusual. Drawing the design of synthetic biology out of the lab and into context, the pilot workshop report hints at how such systems could be made more possible, but only through consideration of complex wide-ranging factors. I see this process as a vital model for good biological design practice.

Time is pressing as we face the spectre of climate change and diminishing fossil fuel supplies. But the rush to build sugar-powered biofuel infrastructure is often underscored by geopolitical or economic pressures around energy, rather than a desire to maintain biodiversity or seek out good design.[29] Perhaps it is our fetish for liquid energy to fuel our lifestyles that obscures better, more disruptive design solutions. With synthetic biology struggling to make scaled-up biofuel production viable, there remains impetus to find alternative models.

This means not only different kinds of energy sources, but different ways of using energy itself. Firing one clay brick emits 1.3 pounds of carbon dioxide, 25,000 bricks require 400 trees for fuel, and 1.23 trillion bricks are made each year.[30] As an alternative to using heat energy to harden bricks, architect Ginger Krieg Dosier's experimental *Better Brick* uses the same microbial-induced

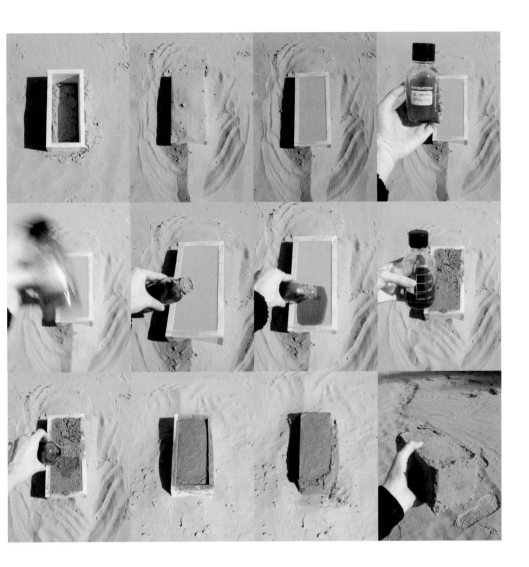

calcite precipitation (MICP) as *BioConcrete*; here it eliminates the need for the oven. Dosier's bricks are made from sand, hardened by a combination of calcium chloride, urea, and bacteria (figure 6.11). There is still much to be resolved in this experimental process: MICP releases ammonia, which Dosier wants to develop ways to capture. But as we seek new energy sources to feed our economy and keep it alive, this kind of disruptive system thinking demonstrates a valuable space for design and synthetic biology to collaborate.

Living Interfaces

In 2009, seven University of Cambridge undergraduates spent their summer designing bacteria to secrete a spectrum of colored pigments for the International Genetically Engineered Machine (iGEM) competition.[31] The team

Figure 6.11
Ginger Krieg Dosier's experimental *Better Brick* process uses microbial-induced calcite precipitation to harden bricks, instead of consuming energy by firing them.

of engineering, biochemistry, physics and genetics students, new to synthetic biology, designed standardized sequences of DNA, BioBricks, and inserted them into *E. coli*. Each BioBrick part contained genes selected from across the living kingdoms that enabled the bacteria to produce a color: red, orange, green, brown, and violet. By combining these with other biological parts, they envisioned that their bacteria—*E. chromi*—could one day be programmed to do useful things, like sense toxins and produce a colored signal.

That summer, at the invitation of Jim Haseloff, designer James King and I joined their two-week "crash course" in synthetic biology, and then worked with the team to find ways to explore the bigger, long-term picture as part of their biological design work. It was an experiment to bring broader questions about the use of synthetic biology into the lab, as the biology was being designed. We ran workshops with the team and their iGEM advisors: in one, we used a timeline not to make predictions, but to understand the agendas— some undesirable—that could shape the use of a living product from the very near future to a century away. Imagining the services, laws, and new social groups that synthetic biology might bring about, rather than the applications themselves, together we considered food additives, patenting issues, personalized medicine, terrorism, and even Google-powered geo-engineering.[32] Such implications are not normally considered at the laboratory bench during the design process. Thoughtful at both the genetic and human scale, conceived with a long-term outlook, their project won iGEM that year. Our collaboration also set a precedent for future work with designers and artists at iGEM by teams around the world, and has inspired engineering teams to follow this way of thinking, too.[33]

One scenario that came out of the *E. chromi* workshops was set in 2039. We speculated that by then, ingesting engineered bacteria would have become culturally acceptable. Probiotic yogurt bought at the supermarket would be drunk for cheap, personalized, disease monitoring. It would contain *E. chromi* bacteria, which would establish a colony in the gut and monitor for the chemical markers that indicate the presence of a variety of diseases, from cancers to colitis to worms. On detection of a disease, the *E. chromi* would start secreting the corresponding colored pigment, producing an easily visible output. Brightly colored poo would signal the need for a visit to the doctor.

We mocked up samples of this fictional technology and carried them to the iGEM competition at MIT in an innocent-looking aluminum briefcase, which we called *The Scatalog* (figure 6.12). We wanted to ask synthetic biologists what a biological computer might really look like. The visual language used to illustrate synthetic biology tends to be clean and shiny, referencing mechanical or digital engineering using cogs, Lego bricks, or computer components. But synthetic biology cannot escape its biological origins. We wanted to address the gap we saw between the materials and the ideas. Biological computers, a new product archetype, may look more like the contents of our briefcase than the neat boxes we understand as computers now. As designers, we saw the gut as an ideal biological computing interface. But is this a future that we want?

In chapter 3, I discussed how by imagining the future, we may unintentionally influence it. Intended as a critical provocation, this project worked more effectively than we could ever have imagined. I often see *The Scatalog* referenced in synthetic biologists' presentations and I have heard of several scientists around the world seeking to make this technology real. Either the fiction was too close to reality, or our intervention did change attitudes. Perhaps most importantly, we brought consideration of social implications into the engineering design process of a living interface. Implications and applications became inseparable.

While some see the *E. chromi Scatalog* as a great design for synthetic biology, there is a huge regulatory and cultural leap between the devices we know and a biological one like this actually being implemented. *The Scatalog* represents a new kind of product archetype, as the boundaries between consumer goods and the human body become indistinguishable. In *The Synthetic Kingdom* (2009), using nonbiological props, I speculated on the physical appearance of consumer products built by biological processes, which bridge this gap between existing materials and unfamiliar future ones. In this fiction, genes sourced from the living kingdoms manufacture food colorants (made more likely later that same year by the *E. chromi* team); light bulbs are filled with bioluminescent bacteria (echoed by Cambridge iGEM 2010's super bright *E. glowli*; see figure I.3 in the Introduction); and biological computers are embedded into visible "microbechips," solid scaffolds that capture bytes stored in colonies. A living "carbon monoxide" sensor has to be cared for, while disposable cups are grown from keratin, the biological plastic that makes our fingernails and hair (figure 6.13). These objects inspire varying levels of disgust: for many, keratin is deemed too close to our own bodies, even rendered in the recognizable form of a recyclable plastic cup.

The Synthetic Kingdom, like *E. chromi*, suggests new locations for design as the fiction investigates the control and spread of biological products. In the second part of the design fiction, products escape factories or are improperly disposed of. The very things we've designed to consume—light bulbs and plastics—colonize us instead, transforming our bodies into places of manufacture. Plastics are molded around the form of the stomach; cyan, magenta, black, and yellow printing inks colonize our teeth as dental plaque and kidney stones bioluminesce. A lung tumor detects carbon monoxide poisoning (figure 6.14). The ultimate synthetic pathology is colonic alchemy, where gold is synthesized from the patient's waste, the impossible achieved. Disease becomes beautiful, synthetic pathologies profitable and even, perhaps, desirable.

The *E. chromi Scatalog* and *The Synthetic Kingdom*'s products and consumer pathologies speculate on what new and unexpected ways that we might see—or choose to see—ourselves and our products in a synthetic biological future. As synthetic biologists seek new design applications, like tumor-killing bacteria, biological hard-drives, or new ways to engineer the human microbiome, finding ways to anticipate and understand the potential implications of living interfaces is of vital importance. Synthetic biology may give us new ways to use and even define the things that we consume as products become indistinguishable from our own biology.

Owning Life

The issue of spread and control is an unavoidable, yet remarkable property of biology. A scientist once told me a story, which captures this unique material trait. A colleague's request for a bacterial strain sample from another lab had been brushed off with a letter of refusal (this was before e-mail). Unperturbed, she simply peeled the stamp—licked in the other lab—off the envelope, correctly suspecting that it would be teeming with bacteria, and successfully cultured out the variety she wanted. This misdemeanor in the name of academic research shows how tricky it can be to contain biology. Designing an organism that pumps out liquid gold in the form of jet fuel also means that it can be stolen, as a physical specimen, or digitally, its DNA "booted-up" elsewhere.

The pressures on the ownership of life forms have shaped biotechnology since the first patents on living things were successfully awarded in the 1980s. Bacteria do not understand borders or follow intellectual property laws, nor do GM crops that spread their pollen. Organisms intentionally released into the wild require both biosafety and intellectual property control. Biotech and "Big Pharma" have been using engineered organisms successfully for decades, protecting their intellectual property

within industrial facilities or by managing farmers. But even these safe-guards can be traversed. Using DIY plant tissue culture techniques, art-ists Georg Tremmel and Shiho Fukuhara "cloned" the Japanese brand Suntory's blue *Moondust* carnation, which the artists describe as "the first commercially available genetically engineered consumer product intended purely for aesthetic consumption."[34] To get them to market, the *Moondust* flowers had been approved as environmentally safe. Tremmel and Fukuhara released their copies to the wild, setting them free in an act of "reverse biopiracy" to make a "Flower Commons" (figure 6.15). The artists' forcible placement of these flowers into the "open-source" domain raises important questions about the control of patenting and ownership of living, reproducing things.

Designer J. Paul Neeley's *Gaia Corporation* is a fictional organization set in 2023 used to satirize these problems inherent in biological intellectual prop-erty.[35] His bloated corporation holds stewardship of the entire intellectual property content of Earth's biology and uses a special formula to calculate licensing costs for biology based on utility, risk, and a special "Gaia Factor."

Figure 6.15
Common Flowers/Flower Commons: A map of Cologne showing sites where Georg Tremmel and Shiho Fukuhara released their "biopirated" copies of *Moondust* carnations, creating a "Flower Commons."

Resolving how property will really be managed is crucial to the future of synthetic biology.

Today, isolated and purified naturally occurring genes are patentable, as are designed and engineered synthetic constructs. At first glance, it seems clearer why synthetic DNA should be owned, as it is more obviously an invention and not a "product of nature." But open-source ideals are foundational to some strands of synthetic biology. The BioBricks Foundation with its "Registry of Standard Biological Parts" argues that innovation will be encouraged if reusable parts are open source and only the constructions built from them patentable. How to establish this legally, and encourage private-sector investment, is an on-going debate.[36]

Copyright and intellectual property are vital to designers and artists too, but new forms of practice are emerging in which protection is sacrificed for innovation. The manifesto for *Open-Source Architecture*, a new movement described in a collaboratively written op-ed in the design magazine *Domus*, offers parallels to synthetic biology.[37] It suggests that the open-source ideology, among other things, encourages "recognising laypeople as design decision-making agents rather than just consumers." Open-source working in many domains is emerging as a tool for the democratization of technology, encouraging mass participation, engagement and discussion, but also innovation. How synthetic biology's interested parties navigate these issues, and how it aligns to a democratic vision, demands substantial open discussion.

Designing Symbiosis

The industrialization of agriculture reveals some of the problems inherent in applying engineering models of simplicity and standardization to natural systems: Monocultures have reduced biodiversity, and the widespread use of chemicals has triggered issues along the complex ecological chain. When Ecuador perceived a potential threat to its biodiversity from multinational corporations involved in industries as diverse as banana farming to gas extraction, it gave nature constitutional rights, including its right to evolution. In 2008, this was ratified by public referendum. Article 71 states that, "Nature or Pachamama, where life is reproduced and exists, has the right to exist, persist, maintain and regenerate its vital cycles, structure, functions and its processes in evolution."[38] Ecuador's constitution charges its people not only with protecting nature, but also with the responsibility to "promote respect towards all the elements that form an ecosystem."

This concept of nonhuman rights may be instructive in shaping an ethical and responsible synthetic biology that is ecologically and culturally inclusive. But what might designing symbiotically with synthetic biology involve? Certainly, considering designs in their ecological context prompts new archetypes: ecosystems of products that work in symbiosis with each other, products symbiotic with existing natural systems, and biotechnologies that are symbiotic with existing platforms.

Since 2009, Srishti College of Art & Design, a small art school in Bangalore, India, has entered a team of art and design students into iGEM each year. Led by Yashas Shetty, the students—many just one year into art school—have not only had to negotiate entering an engineering competition as artists but have also had to define this role. In their first year, the ArtScienceBangalore team experimented with the poetic, designing bacteria that evoked the smell of rain and with it the Indian monsoon. Their *E. coli* secreted geosmin, a volatile microbial metabolite found in the soil and associated with the smell of moist earth, a synthetic trigger of our emotional relationship with "nature."

The 2010 team investigated the concept of a *Synthetic/Post-Natural Ecologies*, illustrating how one modified organism can impact others. They used a common lab transformation: engineered *E. coli* used as a vector to modify *Caenorhabditis elegans* worms. Here the worm is simply reframed: it is described as a biosensor that reveals the presence of modified organisms. The worms' altered state demonstrates the "consequences of interactions between engineered organisms and the 'natural' world."[39] ArtScienceBangalore's 2011 entry took the synthetic ecology out of the lab. In *Searching for the Ubiquitous Genetically Engineered Machine*, the team contemplated a "post-natural" future where BioBricks are an accepted standard and are ubiquitous in the environment (figure 6.16); sampling and mapping the genetic contents in soil could serve as an indicator of human influence.[40] From a molecule that can evoke our relationship with "nature," to a lab-based synthetic ecology, to speculative biological tools that could map the impact of synthetic biology on ecosystems, the ArtScienceBangalore experiments remind us that biological design must be context-aware, at many scales.

Human symbiosis with technology is also significant for the design of biological products and their packaging. In 2006, a team of students from the University of Edinburgh entered iGEM with their prototype *Arsenic Biosensor*.[41] The students engineered bacteria that could indicate unsafe concentrations of arsenic in drinking water, a problem rife in Bangladeshi wells. The biosensor was imagined as a field-testing device. As arsenic triggers the bacteria to change the acidity of the sample, the user could simply test the pH level. This project is typical of iGEM: a real-world problem is used as a narrative driver for lab research, demonstrating how BioBricks could lead to applications.[42]

After iGEM, one member of the team, Kim de Mora, continued to develop the biological design and co-founded a company to spin-out the

Figure 6.16
Students from the 2011
ArtScienceBangalore iGEM
team in the jungle, seeking out
BioBricks in the environment
in their speculative project,
*Searching for the Ubiquitous
Genetically Engineered Machine.*

project, Lumin Sensors. While the arsenic biosensor has come to be iconic of the potential of synthetic biology, translating such aims is another challenge. Start-up projects are vulnerable to the same pitfalls as other emerging technologies, but with its dramatic link between microscopic and macroscopic design, synthetic biology also presents new challenges.

Designing packaging for a biological consumer product prompts questions of reliability, safety, storage, incubation, repeat use, and disposal, all of which impact the biological design of the bacteria themselves. Lumin Sensors collaborated with architect Peter Yeadon and his Rhode Island School of Design students to design prototype packaging for a consumer arsenic biosensor. The social perspective raises further issues bigger than iGEM students can deal with in a short conceptual project. A contaminated well, could mean that a village's girls may be deemed unmarriageable. There are also cultural design differences: The bacteria are designed to incubate around body temperature when the test starts. The American design students responded with ingenious prototypes that could be warmed up in trouser pockets. But since neither saris nor dhotis have pockets, commentators questioned whether one gender might find more reliable ways to store the detector in their clothes, and thus control the technology? Disposal is a major issue with a device containing biological components. In cultures in which resources are scarce, useful-looking things are often deconstructed and saved for re-use. Explaining why the biosensor must be carefully disposed of when a majority of its users do not know about GM triggers a knock-on educational need.

Issues of implementation thwarted this first effort to empower people to test their drinking water cheaply, reliably, and effectively using synthetic biology. Cheaper tests are readily available that do not contain genetically engineered components subject to regulation—extra red tape that complicates a not-for-profit start-up seeking research and development funding. The start-up failed, but scientists from the University of Cambridge and the University of Edinburgh are now developing the biosensor in collaboration with Practical Action, a nongovernmental organization in Nepal with understanding of the day-to-day concerns of the local people. [43]

Finding ways to make design context-aware at every scale is important for the design of real synthetic biological products; otherwise, synthetic biology risks making solutions that are in search of problems, or just more problems that will need to be solved. Designing symbiosis between bacterial communities may present new opportunities as we will see in chapter 17, but designing symbiosis between products, species, ecologies, societies and cultures is unavoidable if synthetic biology is intended to move beyond the confines of the fermenter vat.

Thinking through Things: The Synthetic Aesthetics Residencies

"Cooperation" and "collaboration" are two modes of interdisciplinary practice identified by Paul Rabinow and Gaymon Bennett, anthropologists who have worked with synthetic biologists. Cooperation implies demarcated areas of study with a "defined division of labor," while collaboration is a team effort of discovery.[44] Interdisciplinary collaborations have their own hierarchies: participants can be equals, subordinates, or provocateurs, the work of one group sparking the other. These efforts are fueled by motives ranging from design innovation, critical research, public engagement, to accountability.[45]

The projects throughout this chapter draw on different kinds of interdisciplinary practice between science and design and art. Some were instructive for us as we set up Synthetic Aesthetics; others have appeared during the three-year span of this project and through its developing global network. Synthetic Aesthetics itself has been an experiment in opening up new areas of thought and practice, developing novel language and metaphors, finding roles for art and design in synthetic biology, and innovating interdisciplinary collaboration. Our residents' exchanges were intended as equal as they explored the design of biology. This balance is unusual compared to existing design/ art/science projects, where artists or designers often approach scientists with a predefined question, just as I did when I first e-mailed Jim Haseloff about biological manufacturing.

Faced with an open-ended brief that encouraged speculative research, our residents—artists, designers, and synthetic biologists—were freed from their usual research constraints. Drawing on their range of expertise, their insights address and challenge assumptions at the core of synthetic biology, and their collaborations provide new perspectives on designing biology, which have informed much of the discussion in this chapter. Most importantly, Synthetic Aesthetics has defined a novel space for critical discourse and practice within the biological design process.

As synthetic biology becomes more concerned with making things, its designers need to be aware of how those things sit in the world. The dominant metaphor in the dawning "Age of Biology" has been DNA as the code that controls life. But life is more than DNA. Fetishizing DNA is limiting, and analogies with the digital world prevent us from seeing biology in full. Challenging both synthetic biologists' focus on DNA and the trend for form-based biomimicry in architecture, plant scientist Fernan Federici and architect David Benjamin seek new analogies, questioning synthetic biology's electrical engineering principles (see chapter 7). Manipulating artichoke xylem cells— the plant's water transport system—they prioritize nature's logic over its form. Biology computes complex spatial problems. By harnessing this process, they do more than mimic nature's shapes; they harness its processing power. Their "biological computer" is not just the code that "runs" the cell; it is the whole cell and its context, too, which together generate the logic of the cell's form. Fernan and David's "biological processor" in effect addresses the limits of metaphors. How far can they scale up their "bio logic" before it breaks, as the physics changes? Where are the limits of the relationship between form and function, logic and information?

Working back from form may be familiar to synthetic chemists designing molecules where shape is crucial to function, but in bridging form, information, and time, David and Fernan intentionally and unintentionally touch on issues at the heart of synthetic biology. As they make pattern, form, and context alternative design tools to DNA code and information, they open up new, critical space between two seemingly different disciplines. Their experiments in biomaterials, software, architecture, and bioengineering have informed their individual work, as well as the development of joint teaching, hinting at the new schools of engineering and art and design that the Synthetic Aesthetics project originally hoped to identify.

Biologists Wendell Lim and Reid Williams explored the design process with product designers Will Carey and Adam Reineck (see chapter 9) of IDEO. This residency introduced academic research into a corporate environment and corporate design thinking into the laboratory. The designers' expropriation of the iconic "Hello World" *E. coloroid* plate (illustrated in the

Figure 6.17
Design in our biological future? IDEO logo exposed on a lawn of bacteria by Synthetic Aesthetics residents Will Carey and Adam Reineck (2011).

Introduction chapter; see figure I.1) into their company logo (figure 6.17) hints at the complexities ahead if we permit the instrumentalization of life.

After spending some time in the lab, Will and Adam soon discovered that the promise of easy-to-engineer biology is not yet here, leaving them unable to design the living things they had in mind. They resorted to prototyping design fictions based on their discussions with the scientists to understand the role of designers and the design process in synthetic biology. This collaboration highlights an interesting role for commercial design upstream in science and technology development: Could a firm with the influence of IDEO help establish—and champion—a precedent of good design of biology? The burgeoning relationship between a commercial design company and an academic institution at the early stage in the development of a biotechnology is for me the most intriguing aspect of this project. Both are seeking to establish what a longer-term link would look like. Will we see new departments more akin to MIT's Media Lab as design innovation and scientific research find useful shared territory?

Biochemist Sheref Mansy and artist Sascha Pohflepp worked together on the problem of living machines, making space for philosophical discussion to inform the design process of building life from scratch (see chapter 15). Setting biochemistry within a historical time frame of machines, they use existing things—the horse, the steam engine—to understand better how we might recognize new design subject to the processes of evolution. Without evolution, is a biological design still technically alive? Biologist Christina Agapakis and smell provocateur Sissel Tolaas explore our cultural fear of biology and examine symbiotic design, emphasizing the importance of context in biological design through making challenging objects (see chapter 17) in which we can see ourselves. Biologist Mariana Leguia and composer Chris Chafe allow intangible biological constructs to be experienced in both space and time (see chapter 13).

Bio artist Oron Catts and cyanobacteria biologist Hideo Iwasaki set synthetic biology's endeavors in a far broader timescale, contextualizing synthetic biology's artifacts in the span of all geological time, triggering a thought-provoking investigation into the problematic role of the artist encountering utility, and the scientist, futility (see chapter 11). This discussion, explored further in chapter 12, gives us much to consider when it comes to the strange and occasionally complicated role of artists and designers working within synthetic biology.

Synthetic Aesthetics has shown that there is significant value in bringing design and art "upstream" in the scientific process. It also challenges our expectations: Design can be critical; art, useful. Design's mediating role can be more than a translator between a technology and its consumers. By thinking through "things," many of our residents have made new "useful fictions" (as discussed in chapter 3), working models that highlight the importance of designing biology in context, at all scales, from the molecular to the global, while incorporating biology's complexity, diversity, and temporality. The things they made are the models for us to examine and test, like all of the projects and artworks in this chapter. The parameters that have emerged can inform our understanding of good biological design.

Form may follow function, but we still can decide which functions we want. As we open up a space for artists, designers, and synthetic biologists to examine critical ideas, we return to the idea of design as a medium between ideas and things. And as we think through things together, we are more likely to find a common good in biological design.

Part Three
Synthetic Aesthetics

Logic

Frontispiece
Artichoke Structural Logic.

7.
Bio Logic

David Benjamin
and Fernan Federici

Nature has inspired design and engineering for centuries. For example, engineers have created fiber optics inspired by sea sponges[1] and robots inspired by earthworms.[2] Computer scientists have generated computational methods inspired by the organization of neurons[3] and also by cell distribution in flies.[4] Computer scientists have drawn on biological evolution to develop "evolutionary computing"—in particular "genetic algorithms"—which is widely used for optimization problems, machine learning, and parametric design.

But recent developments in technology and progress in our understanding of biological systems have expanded the range of possibilities for combining design and biology. With synthetic biology—and its enabling technologies—we can now imagine composing new biological systems that never before existed. With new computational tools, we can now use mathematical models to simulate complex biological behavior and explore millions of potential biological designs in the computer.

It is too early to know the exact applications that will be developed in the future, but over the course of our collaboration, we focused on one direction of shared interest. We aimed to go beyond the imitation of biological form and move toward the incorporation of biological function.

In our project, we explored the use of biological systems as direct design and manufacturing tools rather than as formal inspiration. First, we used xylem plant cells as a design tool to render three-dimensional structures in response to different shapes. Second, we explored the use of bacterial patterns as a decentralized manufacturing tool. Our project involves very early explorations, and at this stage it does not propose to offer direct translation from cells and tissues to buildings and structures. But we believe our research will contribute evidence and insights to the new and growing field of biodesign.

Xylem Exoskeletons as a Tool for Biodesign

In plants, xylem vessels are part of the vascular tissue, which is responsible for water and nutrient transport between organs (figure 7.1). These xylem vessels, made of interconnected bars of cellulose and lignin, provide structure for the cell walls to resist the negative pressure of water transport. Xylem structures are rendered in characteristic patterns that seem to vary according to the size and shape of the cells. Thus, the process of pattern formation in xylem cells can be seen as a "morphogenetic" program—it renders form (structural support) in response to the physical conditions of its environment (figure 7.2). This process lacks any external guidance for construction and depends on local molecular interactions. The xylem patterning process is a good example of a decentralized shape-rendering program, so it is a perfect candidate for our aim of using biological systems as design tools. The morphogenetic program of xylem pattern generation can be understood as a "biological design program," and we may be able to harness and apply this program in a completely different context—the context of design.

In this project, we explored several different ways of using the xylem patterning process to find structural solutions to different shapes. To do this, we had to find ways of executing the genetic-encoded process of xylem formation—the biological algorithm—in cells of different shapes and sizes.

In our first approach, we used a technique that allows any cell of a plant (not just the vascular cells) to produce xylem structures when a specific chemical is applied[5] (figure 7.3). More specifically, this enabled us to induce xylem structures—or "run" the biological design program—as a form-solver

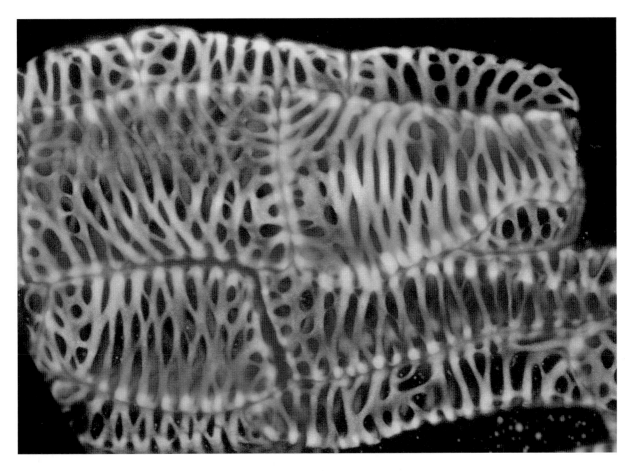

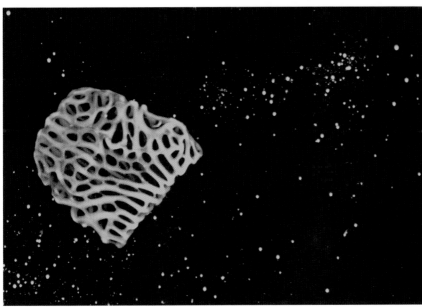

Figure 7.1
Vascular tissue of an
artichoke, showing
interconnected xylem cells.

Figure 7.2
The structure of an isolated
artichoke xylem cell.

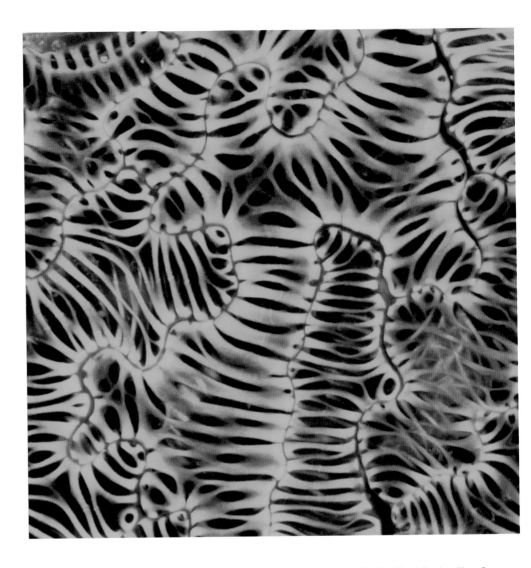

in cells whose shapes were more elaborate than the typical cylindrical cells of plant stems (compare figure 7.3 and figure 7.1). As a demonstration of this approach, we induced xylem patterns in cells of the leaf and solved for the very convoluted shapes of epidermal cells (figure 7.3) and internal rounded mesophyll cells. In these experiments, we obtained structures of consistent patterns despite the large and lobed volume of the cells.

In our second approach, we continued to use the technique of inducing xylem structures in nonvascular cells, but this time we grew isolated cells in microfluidic plates rather than groups of cells in living plants. The use of these plates allowed us to trap cells within custom-designed shapes[6] (figure 7.4). We then induced xylem formation—again running the biological design program under atypical geometric constraints. This allowed us to observe the

Figure 7.3
Chemically induced xylem formation in epidermal cells of an *Arabidopsis thaliana* leaf.

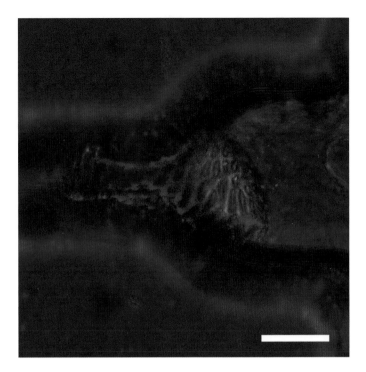

plant's structural solution in response to human-designed "bounding enve-
lopes" (figure 7.4). While our initial bounding envelope was a U-shaped channel, this approach allows for the future exploration of biological response to
many other human-designed envelope shapes.[7]

In our third approach, we used the data collected from the first two
approaches to generate a mathematical model of xylem formation. For each
example cell, we used microscope photos to create a three-dimensional (3D)
computer model. We then situated this 3D model in an array of points and
vectors including a central control line and applied a simple algorithm to
generate a data set describing the geometry of the cell in terms of distances
and angles. At this point, we used a software application called Eureqa to
derive a simple mathematical model that approximated the data set[8] (figure
7.5, top row). With a mathematical model that described xylem formation—a
computer algorithm that roughly matched the results of the biological design
program—we were able to generate new exoskeleton structures in the computer by changing the central control line and the 3D boundary conditions.
We were able to create new designs that were based on the logic of the xylem
cell but that did not require a physical experiment.

In our fourth approach, we explored possibilities for translating the exoskeleton structures from the scale of biological cells to the scale of architecture through the use of advanced computational techniques, including

Figure 7.4
DEX-induced xylem formation
in protoplast from *Arabidopsis
thaliana*, trapped inside a
microfluidic chamber used as
a mold.

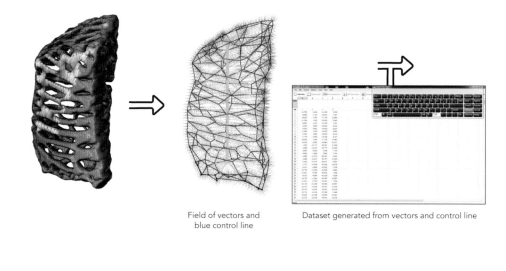

Field of vectors and
blue control line

Dataset generated from vectors and control line

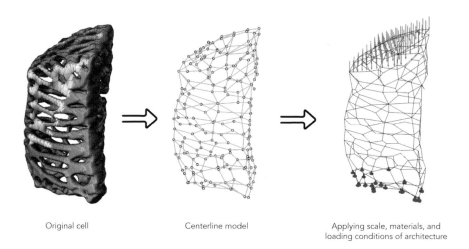

Original cell

Centerline model

Applying scale, materials, and
loading conditions of architecture

Figures 7.5
Applying computation to optimize
xylem structures for scenarios
outside of their natural context.
The top row shows how we derive
a mathematical model of xylem
formation and then run the model
for a new shape. The bottom row
shows how we use evolutionary
computation to optimize the form
for different structural loads.

evolutionary computation and multiobjective optimization (figure 7.5, bottom row). We started with a sample xylem cell and used it to create a simplified 3D model in the computer. We applied architectural scale, materials, and loading conditions to the 3D model. Then, we supposed that our goals for the architecture-scale exoskeleton were to use the least amount of material and achieve the least structural displacement. We ran an automated algorithm to generate, evaluate, and evolve multiple design permutations.[9] With this approach, we derived new design possibilities that extended the biological system and added new requirements of the architectural system. For comparison, we created 3D-printed models of the biological solution and a high-performing, computer-evolved solution (figure 7.6).

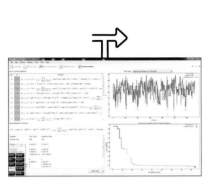

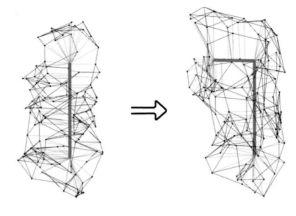

Mathematical equation that approximates the data

Form produced with equation and control line

Alternate form produced with equation and revised control line

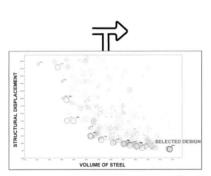

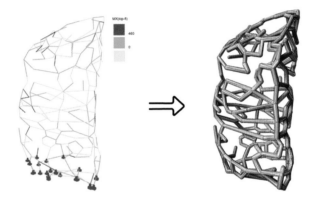

Automated optimization with fitness criteria of least structural displacement and least use of material

Structural performance of fit design

Computationally evolved design

This fourth approach raises interesting questions about whether biological systems should be considered optimal design products. Although each xylem cell represents a good solution for the set of conditions and constraints in which the living system was immersed, this does not necessarily mean that it is the best structural solution for all scales, materials, and loads. A given xylem structure extracted from a living cell represents an optimized solution for that specific shape and size. It is perfect given the type of material available for its construction, the viscosity of tissues, the conditions of nutrient transport and permeability, and the mechanical stability provided by surrounding cells. Yet in most cases, direct translation of these tiny natural structures to a larger human construction would be suboptimal—in large

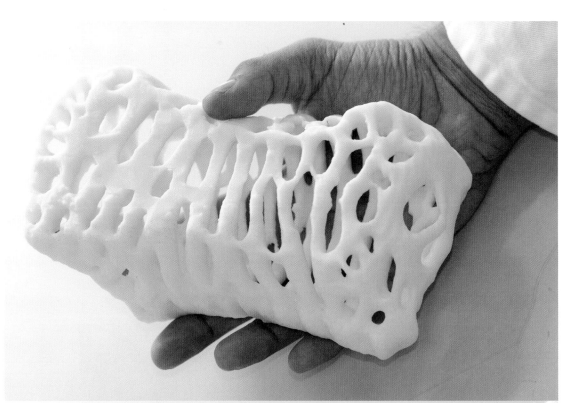

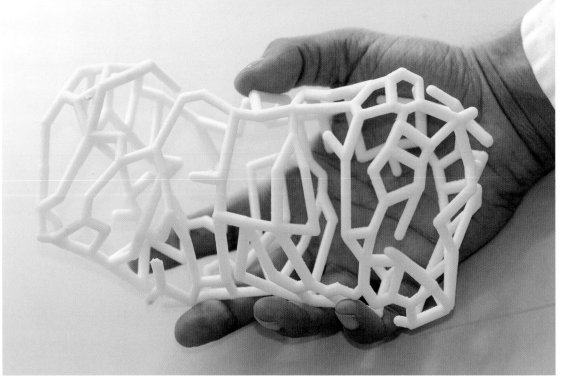

part because moving from cells at the scale of micrometers to architecture at the scale of meters involves a million-fold shift in scale. Therefore, the use of computational optimization seems to be promising for translating scale and obtaining a high-performing design solution.

Although we investigated the problem of translating scale, the aim of our project is not specifically to use xylem cells to design buildings, but rather to explore the general potential for using biological systems as design tools. The xylem system is a very visual and immediate example, but more directly useful applications could be developed in the future with other living systems.

In summary, each of our four approaches has different strengths and limitations, and our development of all of them is very preliminary. Yet for each of the approaches, we have demonstrated the complete workflow with one or more example designs, and we believe that they all represent promising directions for further research and refinement toward the goal of using biological systems as design tools.

More broadly, this new form of "cooperation" between a human designer and a living cell allows us to produce designs that a human alone—or even a human and a computer—could never create. Here we are not simply replicating the form of nature—we aim to replicate the logic of nature. This new design process involves designing with shifting and sometimes unknowable forces, and designing dynamic relationships rather than designing fixed forms. These are huge transformations from the typical design process, but they are fitting for the dynamic and unstable world in which we now live.

Bacterial Biofilm Patterns as a Tool for Biofabrication

Bacteria naturally produce complex spatial patterns (figure 7.7), often involving the interaction of different species of bacteria. They also create biofilms—they manufacture tangible physical materials—under specific conditions. For example, bacteria can create rigid, brick-like material through the process of microbial-induced calcite precipitation,[10] and they can also create flexible, fabric-like material through the process of microbial cellulose production.[11] In this project, we explored the combination of these processes—pattern formation, manufacture of rigid material, manufacture of flexible material—in order to generate novel composite architectural materials. In contrast to applying architectural technologies such as digital fabrication and computer numerical control (CNC) manufacturing machines with a fixed and predetermined physical output, we experimented with applying manipulated biological systems for a bottom-up approach to fabrication.

Our exploration aimed to conduct corresponding tests in software and in the lab. For testing in software, we developed a workflow and user interface that allows for the exploration of a range of different bacterial colonies

Figure 7.6
Three-dimensional printed prototypes comparing the cell and the logic of structure.

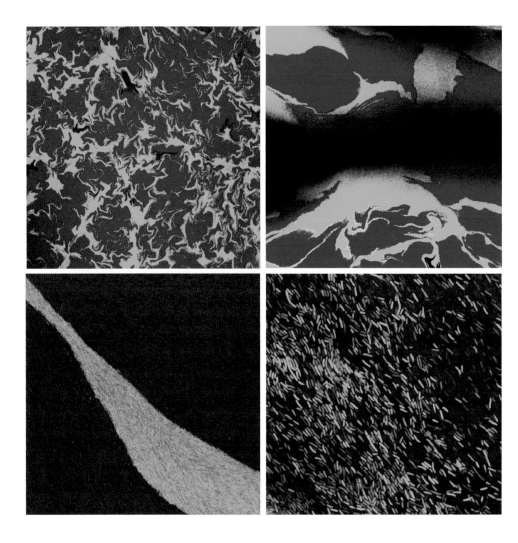

Figure 7.7
Escherichia coli colonies
expressing red and green
fluorescent proteins in
a biofilm, shown at increasing
levels of magnification.

and the different sheets of material that they manufacture. Our software allowed a user to conduct virtual experiments with a few basic steps. First, the user adjusted the conditions of a Petri dish and the features of two types of bacteria growing in the dish. Then, the user simulated bacterial patterning, bacterial deposition of material, and the structural properties of the resulting sheet of material. Finally, the user ran the software to automatically explore thousands of design possibilities by varying the inputs (properties of environment and bacteria) and searching for better outputs (structural performance) (figure 7.8). In this process, the software was used to design new composite building materials through a novel manufacturing process of bacterial patterning and deposition.[12]

While testing in software allows for the low-resolution analysis of tens of thousands of design permutations, testing in lab experiments allows for

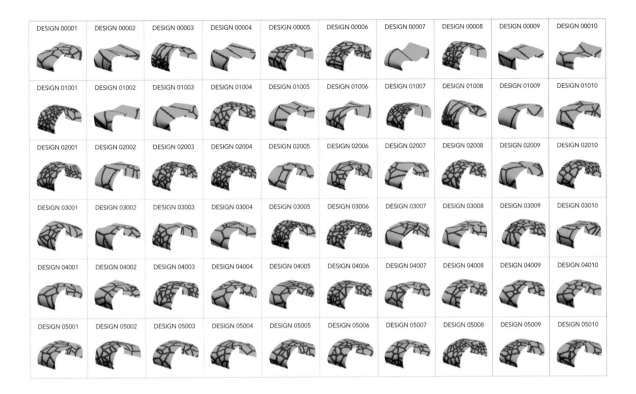

the high-resolution testing of a few dozen design permutations. In addition to our computational process, we experimented with the implementation of bacterial patterning mechanisms in the lab (figure 7.7).

This line of research suggests several transformational possibilities for architectural manufacturing and construction. Building materials generated through bacterial manufacturing may involve higher performance, lower cost, fewer carbon emissions, and less waste than typical building materials. Bacteria may be able to produce composite building materials with microscopic arrangement of materials that would not be possible through other manufacturing processes. Because many bacteria consume a variety of sugars, the raw materials required for this manufacturing process may signal a transition to a "glucose economy" that eventually replaces the petroleum economy and alters many aspects of global trade and transportation. In other words, bacterial manufacturing may enable a new movement for "local materials" similar to the current movement for "local food." And as bacteria are self-replicating organisms, it may be possible to devise new systems of building, sending only a small test tube or even a computer file of DNA code to the construction site rather than sending premanufactured building elements. This final example could be especially useful in extreme environments such as Mars, where the cost of shipping typical building materials is prohibitively high.

Figure 7.8
Computational evolution of composite sheets made by the logic of bacterial patterning. Green regions indicate flexible material, and red regions are rigid.

Research Directions and Educational Models

Biodesign and biofabrication suggest new directions for the combination of biological systems and architecture. Although there is a long history of using natural materials in construction and imitating natural biological forms in architecture, synthetic biology offers the opportunity for direct synthesis of form and shapes through engineered morphogenetic instructions. These engineered biological systems promise to open up a new field of applications. In other words, when designers imitate natural forms or directly translate patterns of biological organization, they are exploiting only a very limited number of permutations in the potential design space (the space of all the potential organizations of the same elements). These naturally occurring permutations are constrained by evolutionary pressure and also by other limiting factors such as scale and materials. Yet there might be a number of permutations that never thrived in nature and therefore were never before considered for use in architectural design. By using new engineering principles of synthetic biology, new computer modeling techniques, and new technologies of DNA synthesis, designers might be able to explore new regions of design, and they may be able to create new design products that grow, adapt, and replicate.

A new framework of biodesign could emerge in which the practitioners explore the development of synthetic biological systems that are free of unnecessary adaptation constraints but still exploit the robustness of natural systems. There are already good examples of using biological systems as direct problem solvers.[13–15] Our project aims to contribute to this framework of biodesign and more particularly to the idea of combining the nonlinearity and multiparallel information processing capabilities of natural systems (the xylem program) with the possibility of direct synthesis of material as an output (bacterial patterning and material deposition).

This new framework of biodesign could also lead to new schools that combine architecture and synthetic biology. As with our own research, these schools could build on two existing educational platforms. First, they could use the International Genetically Engineered Machine (iGEM) competition and its Registry of Standard Biological Parts as a platform for exploring new biological systems for specific applications. Second, they could apply the educational model of design studios in architecture schools as a platform for intensive investigation of design possibilities through a combination of creative intuition and systematic experimentation. Over the past 2 years, we have explored our two research projects in the design studio with master of architecture students at the Columbia University Graduate School of Architecture, Planning, and Preservation, and we intend to continue and expand the use of this teaching and research.

8.
Scale and Metaphor

Pablo Schyfter

David Benjamin and Fernan Federici's collaboration has been an exercise in finding new ways to work across disciplines. Importantly, what this pair managed to accomplish is more than just an exercise in trading knowledge. That is to say, David and Fernan didn't simply pass along to each other specialized expertise. Without a doubt, each participant spent time teaching the other, and each did leave the collaboration having gained new understanding. However, this trade in know-how captures only a small fraction of their work. The back-and-forth of teaching and learning does not encompass what was a truly collaborative, joint effort at developing something new from the union of two disciplines.

David and Fernan made use of resources from architecture and synthetic biology to construct a set of tools and techniques unique to their collaboration. In developing and putting to use these tools, the pair has shed light on important aspects of design and synthetic biology. Among other things, their partnership challenges us to explore how metaphors and scale shape the field's work with engineering and living things.

Marrying Architecture and Biology

David and Fernan's project brings into dialogue and partnership two very different fields—biological science and architecture. Fernan works at the University of Cambridge as a researcher in plant biology. He makes use of synthetic biology's techniques and technologies to study patterns of growth and physical change in plants. As part of this work, Fernan has developed great expertise in microscopy and imaging techniques—skills that played an important role in his work for the Synthetic Aesthetics project. David is an architect and also teaches at Columbia University. One of his professional interests has been the use of computational optimization in producing building designs—using computers to arrive at different solutions to design problems in architecture. Clearly, the traditions, canons of knowledge, routine practices, and values of architecture and biological science are markedly different. Given their divergent aims, it would make little sense for these two disciplines to share a great deal of common ground. Nonetheless, David and Fernan found many commonalities and shared interests in thinking about design and synthetic biology.

David's work in architecture has two key ingredients that made for compelling joint work with Fernan: his use of computer optimization software for design and his strong stance against so-called biomimicry. As David understands it, biomimicry is the copying of form from nature. For instance, an architect may design a building to look like a leaf or a shell or, in the case of the Kunsthaus Graz, an organ (figure 8.1). If the building's design draws no other inspiration or lesson from the way leaves or shells are shaped or grow in nature, the design does nothing more than mimic form. It copies shape with no thought given to the underlying causes of that shape. Biomimicry is, according to David, a simplistic and superficial way to bring biology into architectural practice. As he often said, it is about form, but not logic. Yes, we can copy shapes, but there is nothing profound about doing only that. Something deeper and more meaningful is possible. David stressed that learning how and why natural things grow as they do may teach the architect about new and better ways of designing structures.

Fernan shares with David an interest in the alluring complexity of living things. His work at Cambridge focuses on "Turing patterns." Examples of

Turing patterns include such things as a tiger's stripes or the arrangement of a tree's branches and leaves (figure 8.2). All animals or plants of a species will have these patterns, but no two will have patterns that develop in the exact same way. Two tigers of the same species will both have stripes, but those stripes are not identically formed. Trees have branches and leaves, but no two trees have identically formed patterns of branches and leaves. Turing patterns, Fernan argues, point to the complexity and unpredictability of biological things. Living nature is not straightforward. This perspective is one that both David and Fernan hope to bring to the world of synthetic biology. Rather than work to reduce or eliminate complexity, one of synthetic biology's stated goals, the field might instead find ways to celebrate, learn from, and harness unpredictability.

David and Fernan's collaborative work began in the world of computation. When they first started, computation was a way of bringing their different fields into some kind of partnership. Software tools helped to bridge the

Figure 8.1
Architects Peter Cook and Colin Fournier's Kunsthaus Graz (2003) is a striking example of biomimicry in architecture, resembling a biological organ of some kind.

Figure 8.2
Examples of Turing patterns
created using a computer.
In nature, broadly similar patterns
will be present in two members of
a species, but those patterns will
never be identical.

disciplinary gap between the domains of architecture and biological science. The idea of computation carried on as the foundation of their project and made possible their critical contribution to the Synthetic Aesthetics project's exploration of design and synthetic biology. Nonetheless, it started as a way to construct links: a way to connect Fernan's and David's day-to-day work.

An important facet of Fernan's research involves microscopic imaging, and his photographs have gained wide attention for their quality. For their joint work, Fernan and David used confocal microscopy. This technique produces images consisting of horizontal "slices" of objects, say plant cells. These "slices" can then be used to construct three-dimensional digital models of those cells. In architecture, three-dimensional modeling is a widely used tool. Here was the first suggestion of a connection; the partners both used computer modeling of structures. As they explored this similarity further, they discovered commonalities in the software tools they each used. The next step was the result of inspired curiosity: What if we use the tools of biology to make images, but then use the tools of architecture to study the models?

Using Fernan's images, David developed digital models of xylem cells. Xylem cells make for fascinating objects of study and modeling because they produce a type of exoskeleton, a rigid lattice-like structure, around themselves. The pair imaged these microscopic structures and then used the photographs to construct digital models. These models were then subjected to computational optimization. That is, David used the kind of software employed by architects to solve construction problems in order to test out how biology might try to solve its own growth problems. The laboratory practice of microscopy was transformed into the studio practice of structure optimization. Natural growth and computerized construction were placed face-to-face and set into conversation.

The imaging tools of biology and the optimization software of architecture were made to serve different ends, and they do not connect without difficulty. Bringing them together involved moving past just trading know-how in order to develop something that is uniquely a result of this particular collaboration. Nonetheless, shared tools were only the beginning. David and Fernan put those tools to use in building a shared perspective on synthetic biology.

"Biocomputing" and the "Logic of Biology"

The team's technical accomplishment in linking microscopy and architectural optimization software is an interesting and innovative result of their collaboration, but more important are the critical perspectives introduced as a consequence of that effort. Beyond the ability to couple tools from biology and architecture, what traffic of ideas, practices, positions, and debates did the project make possible? What do we learn about synthetic biology? About nature? About design?

Having developed a set of common tools for their project, David and Fernan made of their collaboration an exercise in using those tools. They used the tools in a series of joint experiments on xylem cells and also used them to build a common language for thinking about design and synthetic biology. As they built and used their software, they also developed concepts and terms. Two of these—the notions of "biocomputing" and "the logic of biology"—stand out as particularly important.

Perhaps unsurprisingly, making use of computational tools from biology and architecture quickly led the team to question how computation bears upon the things of living nature. Many previous and ongoing projects in synthetic biology aim to develop biological equivalents to standard components used in electrical and computer systems engineering: logic gates,[1] switches,[2] oscillators,[3] and memory devices[4] are examples. These constructs draw inspiration from electronics but are biological in makeup. David and Fernan were compelled to ask: What is "biocomputing"? At one extreme, the term might simply refer to good, old-fashioned computation now done with biological materials; that is, somehow changing living things so that they can do what our existing computers can do. In all likelihood, this would involve a considerable amount of work in shaping biological things to behave as expected, but it would involve almost no change to our understanding of what computing is. The "bio" in "biocomputing" would refer only to the material being used. Much of David and Fernan's work has been about exploring an alternative understanding of "biocomputing"—one based not in human constructs, but rather biological behavior. For them, the term signifies more than just existing computation done with biology; it suggests a distinctive type of computing uniquely of biological nature.

David and Fernan put their understanding of "biocomputing" to work in a series of experiments. They set about to test how organisms react to simple problems posed by their environment. They wanted to explore how those organisms "compute" solutions. At Cambridge, the two subjected xylem cells to particular physical constraints and then observed how the cells grew within the constrained space. In parallel, David and Fernan used an architectural software package to postulate what the final form of the cells would be: How would the cells grow if they followed the logic of human-built structures? When the two were compared, the efficacy of the computer model in predicting the cells' behavior could be determined. In the most straightforward terms, these experiments allowed Fernan and David to improve the quality of their computational tools gradually. More importantly, these experiments made evident some unique ways in which biology arrives at solutions to its problems. The team came to call this "the logic of biology."

As the pair understands them, "biocomputing" and "the logic of biology" are interlinked phenomena. In one sense, "biocomputing" is "the logic of biology" in action. In experimenting with how biology might compute in distinctive ways, David and Fernan were in search of "the logic of biology." In discussing this "logic of biology," the pair puts forward a compelling argument about what "biocomputing" is and how synthetic biology might harness

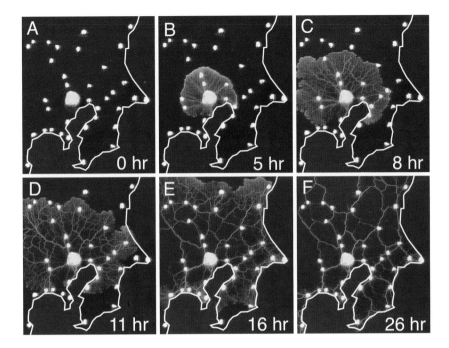

it. This argument is a compelling counter-narrative to the position that underlies much "biocomputing" research currently under development in the field. Rather than impose a digital logic on biological organisms, David and Fernan set out to learn from biology how it "computes" answers to certain problems by growing in certain ways. The pair often used an interesting example of research with slime mold to illustrate this point. Researchers set up an environment for slime mold growth in which food sources mapped onto major population centers in Tokyo[5] (figure 8.3). Slime mold was introduced, and its growth mirrored the configuration of Tokyo's mass transit system. Here, slime mold "computed" the solution to a human problem. Instead of viewing the biological as material to be shaped by human hands, the team presented living things as sources for inspiration and learning.

Scale and Metaphor

Within the scope of this book's critical look at synthetic biology, the importance of David and Fernan's project has less to do with enabling new types of architectural practice or new tools for the study of biological phenomena. Both of these are admirable achievements, but the team's work matters more for its ability to raise questions and provoke discussion. In translating across fields and developing computational tools, the team posits questions about what underlies current practice in synthetic biology. Many of these questions center on synthetic biology's implicit claim—at best unproven, at worst

Figure 8.3
Scientists used slime mold, *Physarum polycephalum*, to "compute" a solution to mass transit in Tokyo. This is an example of harnessing the logic of biology in order to resolve a human problem.

false—that what makes sense in the realm of technology applies equally well to the things of living nature. The team's work with biocomputing and the logic of biology engages with what computing in a biological realm entails and poses a challenge to the position that a digital logic makes sense when applied to living things. More broadly, the team brings into focus two important issues about synthetic biology's ongoing work: scale and metaphor.

The use of analogies with electronics in making with biological things rests on the presupposition that moving through scales is not problematic. That is, what makes sense at the human scale of laptops is just as valid at the microscopic level of nucleic acids—the "logic" of silicon and the "logic" of DNA are basically interchangeable. Notably, David and Fernan found the reverse of this assumption at work in some architecture. That is, the pair found examples of architects assuming that microscopic "logic" works at the scale of human buildings. Problems with translation can be found in both of the residents' disciplines. It was not lost on the pair that such attempts at translating across a tremendous difference in scale—from the cellular to the architectural—might well fail. David's dislike of so-called biomimicry in architecture is relevant here. Remember that biomimicry involves copying the shape of natural things in making buildings, without giving thought to why that shape exists as it does; for example, constructing a building that looks like a leaf or shell, but in all other respects draws no lesson from how leaves and shells are grown and shaped in nature. Similarly, both David and Fernan find synthetic biology's scaling-down of such things as electronic circuits a suspect practice. Doing so pays no attention to why the form of circuits may make sense only when those circuits are built with metals and plastic. Synthetic biology often positions itself as a remarkable new form of engineering because it works with a type of material never before engineered in a systematic way. If this is true, shouldn't the field seriously question the applicability of existing technological logic to this new realm? Working with a sensitivity to scale demands addressing this issue. Scale, then, involves more than just size. While size is a key starting point, scale is about the act of translating across different materials, behaviors, contexts, and "logics." To ask about scale is to question, for instance, the sensibility of treating human achievements in electronics as applicable across the board of genetic phenomena.

Thinking about the logic of biology and the role played by scale quickly leads to the many comparisons, analogies, and metaphors that shape how synthetic biology looks to the living world and what it aims to accomplish. Metaphors are a mundane part of everyday language[6]; nobody is literally the "black sheep of the family," and never does it actually "rain cats and dogs." Nonetheless, using those expressions communicates ideas in a way that is shared by other people in the same community; as long as one is part of that community, its metaphors will make sense and will be useful. As a particular type of community, synthetic biology also has particular metaphors. Some of these are off-handed everyday expressions, and others are used as part of the field's work. As I discussed in chapter 5 on nature and design, metaphors can work as shortcuts and placeholders. To say that DNA is "read" or that it is a "program" is to make use of metaphors. DNA is a chemical structure; it is not a text or software. In both examples, the literary trick is a useful one. It works as shorthand for much more extensive and sophisticated claims about what DNA is and what happens to it. In this regard, metaphors are simply useful stand-in expressions. However, the usefulness of metaphors is accompanied by somewhat more problematic results. When researchers in synthetic biology talk or write about making "circuits" using DNA, they are implicitly presenting cells as computers that process information. That is, they are subscribing to the type of problematic reasoning that imposes an electronic logic onto the stuff of biology. Referring to a "circuit" may be dismissed as a useful shortcut, but language has very real, material consequences for our relationship with other living things. Engineering metaphors in synthetic biology shape how researchers think about and work with living things. Think and talk about biology as a system of electrical components, and that is how you will engage with it. In this sense, metaphors can serve as gateways to the type of problems with scale that David and Fernan's work with biocomputing explores.

The pair's work identifies a type of problematic practice found throughout synthetic biology. Researchers translate across scale without considering the problems with doing so; complexity is brushed aside in favor of simplistic assumptions. These are consequences of an overarching approach to synthetic biology: the imposition of engineering logic onto the stuff of living nature. Not only may this be a counterproductive way to design with living things, it also excludes any and all things that don't fit within the engineering vision. Complexity and unpredictability, two important facets of living things discussed by David and Fernan, are among such excluded phenomena. These residents enable a telling perspective on this facet of synthetic biology. In thinking through what biocomputing means and trying to provide an alternative definition, the pair illustrates how imposing metaphors that exclude key characteristics of living things leads to a simplistic understanding of those things and ignores many of the properties that make biology a compelling domain to study in the first place. Alternatively, finding where metaphors fail and when differences in scale matter can serve to undermine dubious assumptions. This is precisely the reason why David and Fernan sought to reimagine the concept of biocomputing. As long as the idea of computation is drawn directly from the realm of electronics, it will be an analogy imposed on biology. If it becomes a concept that can have a different, unique manifestation in the world of biology, "computation" can serve as a tool with which to understand living things and learn how to harness their distinctive properties.

The issue of scale concerns the act of moving between different domains of size, material, behaviors, and contexts. As David and Fernan demonstrate, it is about varying "logics." Translations across scale can lead to superficial, limited results such as biomimicry. Alternatively, if differences are taken seriously, moving across scales can be a challenge to satisfy and a chance to learn. David and Fernan's work encourages synthetic biology to think about what makes the most sense for designing with living things—a "logic" from the world of inanimate objects or one from the things of biology itself. A place to begin encouraging this more worthwhile perspective is language: Metaphors should be opportunities to think about the relevance of engineering to biology, rather than the result of assumptions made about its obvious applicability. Until metaphors become questions to study instead of given realities, synthetic biology will fail to design *with* biology. Rather, it will continue to design *on* biology.

Process

9.
Living among Living Things

Will Carey, Wendell Lim,
Adam Reineck, and Reid Williams

What might a world look like in which engineered biological organisms are an evident feature of our everyday lives and are able to play a valuable role in solving essential human challenges? What could we achieve by using biological materials—engineered microbe, plant, or animal cells and tissues—as a substrate for design? What would it mean to "design nature" in this way?

These questions animated a series of activities and discussions involving the participants in our Synthetic Aesthetics research between synthetic biologists Wendell Lim and Reid Williams of the University of California, San Francisco, and industrial designers Will Carey and Adam Reineck of the innovation and design consultancy IDEO.[1] Through cross-disciplinary collaboration, we aimed to explore how the nascent science of synthetic biology could radically change both the products we use on a daily basis and the process by which they come about.

In this chapter, we describe our collaboration, its challenges, and the tangible outcomes we created, and we reflect upon what the experience taught us, as designers and scientists, about new possibilities in our own and each other's disciplines.

Jumping in at the Deep End

Few collaborative endeavors between designers and laboratory scientists take place, save for the design of the lab equipment or space in which scientists work. Theoretically and methodologically, the two professions seem worlds apart. From our very first meeting, it was apparent that the two universes of industrial design and synthetic biology have two separate languages and sets of collaborators. Hence, it was important to take time to understand and become conversant in each other's ways of thinking and doing, including the cultural systems within which we operate. Along the journey, we regularly brought in experts and advisors from our respective professional networks to engage with the project and help bring fresh eyes to the challenges we wanted to explore together. Extending the project beyond a collaboration between one designer and one scientist presented an opportunity to explore and collaborate in innovative ways and, in the process, to discover how design and science as professions might work more closely in the future.

The microscopic landscape of cells and molecules is largely unfamiliar territory for designers accustomed to observing and creating at a human scale. The extreme precision and rigor needed to study and manipulate cell signaling pathways was striking to the designers, primarily from a temporal standpoint. Designers often use processes that are rough, rapid, and right—such as quick prototyping from paper, modeling foam, or software and hardware platforms like Arduino—to refine and build new ideas. Such iterative prototyping processes enable ideas to be tested quickly before committing to moving in a particular direction. The immersion in experiments that modified the functions of *Saccharomyces cerevisiae* yeast and *Escherichia coli* bacteria, for instance, brought about a recognition that many of our ideas would, at least for the time being, remain unrealized. The time, knowledge, and effort required to prepare a single experiment were sadly beyond the scope of this project. However, close contact between the scientists and designers ensured that the ideas we generated do represent genuine scientific possibilities.

Our aim was not so much to produce concrete results or usable products as to initiate an open discussion about the direction that synthetic biology

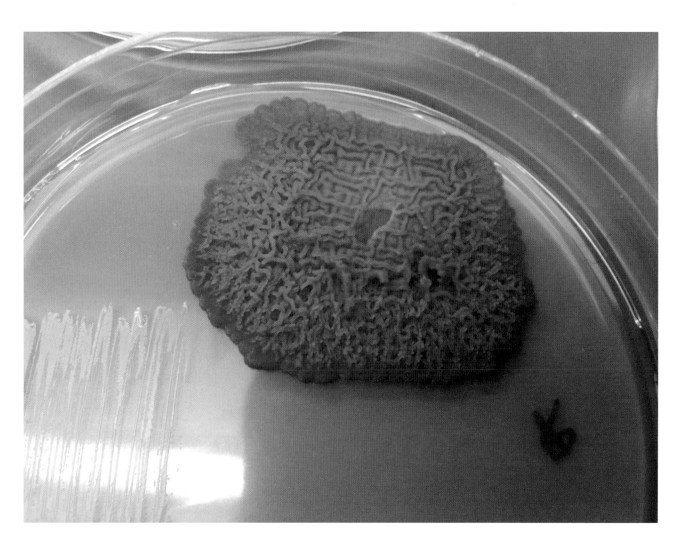

might take in the future. We wanted to consider how the design process could generate new ideas that made use of synthetic biology while remaining unconstrained by the usual practicalities and complexities of the scientific process (figure 9.1).

Creating Life/Giving Rise to New Ideas

After the design team had spent two weeks in the Lim lab getting a feel for basic lab skills that ranged from pipetting tiny volumes of liquid to copying pieces of DNA, we began our discussions on the future of synthetic biology together.

The designers organized a series of creative, knowledge-sharing lectures and discussion sessions for the scientists at the design studio, culminating in a one-day "deep dive" workshop (figure 9.2). This workshop placed members

Figure 9.1
Inspiration for the material structure of our *Packaging That Creates Its Contents* concept. A mutation in a yeast metabolic gene causes it to accumulate a red-colored chemical.

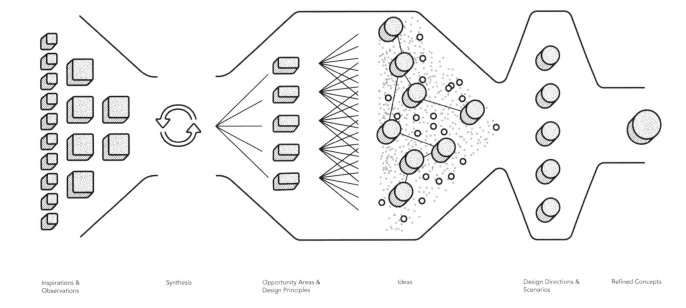

Inspirations & Observations Synthesis Opportunity Areas & Design Principles Ideas Design Directions & Scenarios Refined Concepts

of the Lim lab in the position of collaborative partners with IDEO, participating in the design process from initial brief through to rough prototype and concluding with the presentation of a selection of ideas. The goal was to spend a day thinking as a group about synthetic biology not simply as a field of scientific research but as part of a broader human context. The IDEO designers wanted to learn what the impact of synthetic biology could be on people's everyday lives and how it could revolutionize the industries and products we use and interact with on a daily basis.

To explore this, we ventured into our local Whole Foods supermarket in downtown Palo Alto, California. Each participant had to choose an everyday item priced under $10 that synthetic biology could influence in the next 20 years.

We came back with items that included probiotic drinks, food, laundry detergents, and skin-care products that served as the conversation starter for our design "deep dive." Together, the designers and scientists brainstormed, sketching and making paper prototypes of future products. We split into small groups to develop selected ideas and presented them to the larger group. In the final roundup, we identified a common theme: "living among living things." This is an imagined world where the everyday products with which we coexist are alive in ways that are not always recognizable at first glance. Two possible scenarios from this future, illustrated here as design concepts, are intended to inspire dialogue and debate about the social, economic, and ethical implications of this emerging field of science and the living products that it might make (figure 9.3).

Figure 9.2
IDEO design process explaining the moments of divergent and convergent thinking that lead to final designs.

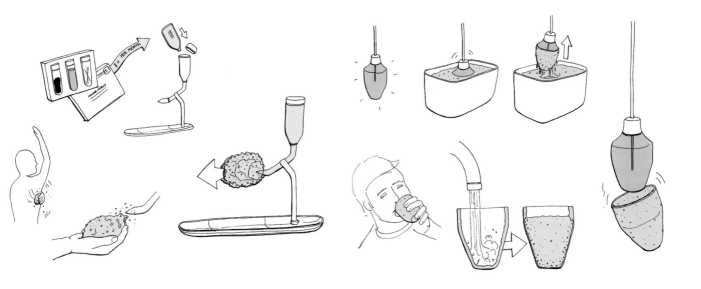

Future Concepts

As designers, we are passionate about envisioning and making physical things that shape our environment. What excited us most as we learned about synthetic biology was the potential for humans to develop a closer relationship with other species, both inside us and around us. Our culture's current mindset encourages us to isolate ourselves from the natural world; typically, we shelter ourselves from the elements, sterilize ourselves and our environments, and do all we can to ensure we don't come into contact with germs and bacteria. The tangible design concepts we have developed, illustrated through renderings, represent a way toward a greater awareness of closer, mutually beneficial relationships with other living things.

Personal Microbial Culture

What if we could nurture organisms that could be tailored to meet our body's needs? We are understanding more and more the complex symbiotic relationships between human bodies and their resident microbial populations, including those in the gut and on the skin. Probiotic drinks and the importance of "good bacteria" in the digestive system are increasingly familiar. How could a similar concept be applied to improve the health and appearance of the skin?

The skin's appearance is based on a complex series of interacting variables, including an individual's genetic makeup, the skin's microbial flora, age, exposure to ultraviolet light, and externally applied skin-care products. We discussed how we might influence the condition of the skin by manipulating

Figure 9.3
Concept sketch for *Living among Living Things*.

microbial populations through the application of signaling molecules—the chemicals used to transmit messages between cells. Based on an understanding of these interactions, could we create personalized skin-care products using a sample of the individual's genome sequence and skin flora?

We envisioned a personalized skin-care product, secreted by a living population of microbes housed in a purpose-built container. These engineered, synthetic organisms feed off the cotton balls that are used to apply them to the skin by producing cellulases: enzymes that digest cotton to produce glucose, the cell's primary source of energy. The organism could also produce fragrance, soap, and oil molecules, vitamins, and signaling molecules in a combination that is most appropriate for the individual's skin type (figures 9.4 and 9.5).

Packaging That Creates Its Contents

What if we could train bacteria to grow into physical products, creating a programmable, biological manufacturing process? Synthetic biology is already producing new types of chemicals and fuels that slot into existing industrial supply chains, hoping to replace those derived from costly extraction processes.

In this scenario, we conceived a synthetic organism that has the capacity to form a more structurally complex material, which constitutes both product and packaging. Upon exposure to a specific wavelength of light, this engineered bacterial population would form a rigid, waterproof cup. During shipping and storage, these light-molded cups remain alive, but dormant.

When exposed to water, these beneficial microorganisms hydrate and begin to produce fragrance and flavor compounds, creating, in this instance, a nutritious and refreshing probiotic drink. After several uses, the cup starts to degrade and is ready to compost (figure 9.6).

Design's Cultural Influence

Never before has the language of design been such common currency within the popular media, resulting in a general public more aware of design as a practice and a profession. To create desirable products and services in the competitive modern marketplace, most of the world's major corporations now recognize the imperative to give designers a prominent seat at the table. Corporations such as Apple have placed an aesthetic sensibility at the heart of their approach in a way that has generated prodigious sales and brand loyalty. Design is no longer an added extra, the selection of surfaces, fixtures, or fittings, but a way of thinking that is fully integrated into the way organizations, from charities to large multinationals, do business.

Every thoughtful designer aims to push the limits of design by applying skill and aesthetic sensibilities to improving the quality of our lives and the objects and systems that surround us. Designers are contributing to culture

Figure 9.4
Envisioning taking a skin sample for *Personal Microbial Culture*. The skin sample contains the user's unique microbial flora.

Figure 9.5
After a skin sample is taken, a microbial culture would be created that produces skin-care products customized for the user. The culture feeds on used cotton balls.

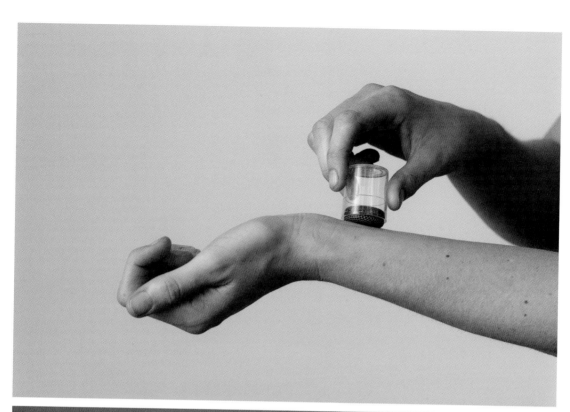

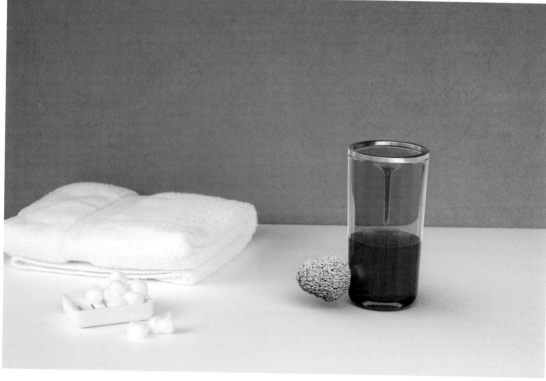

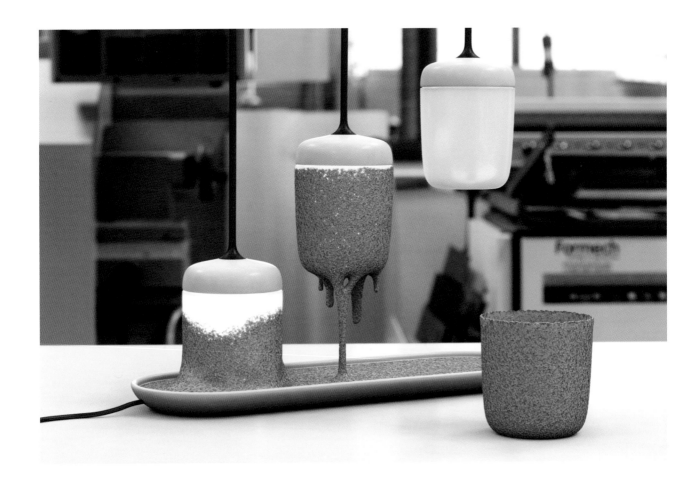

and creating visions that lead to the future. Often, this implies a compulsion to engage with the latest ideas and newest technologies emerging at the cutting edge of science and engineering.

The growing understanding of design is allowing it to move further "upstream," to assume a valuable role in creating a vision and provoking debate about how the technologies of tomorrow can be used. Design is becoming more than simply a means to utilize today's technologies for today's purposes. Designers can present a powerful vision of the future, facilitating visceral and tangible forms of engagement with innovation and possibility, and exposing their audience to new ways of being and interacting in the world.

Creativity in Science

Science tends to be viewed as one of the most rigorous and logical of human endeavors. Yet this popular impression of the practice of science is based on how it is presented to us, rather than how it is practiced in reality. The great French molecular biologist and 1965 Nobel laureate, François Jacob, wrote

Figure 9.6
Imagining bacteria-based manufacturing. Microbes form a layer around a mold. Light evokes a material change in the bacteria, causing them to enter a dormant state in preparation for shipping. A probiotic drink could be activated by water, which triggers dormant microbes that have formed the container itself.

of the two faces of science: "day science," the official science of articles, seminars, and textbooks, and "night science," the world of confusion, doubt, sweat, and inspiration. He suggested that in reality, path-breaking scientific research emerges from "a jumble of untidy efforts, of attempts born of a passion to understand." These efforts are then written up in such a way as to depersonalize the achievement and "replace the real order of events and discoveries by what would have been the logical order, the order that should have been followed had the conclusion been known from the start."[2]

When a scientific project is completed, and some nugget of knowledge has been gained and verified, it is much more effective to communicate this knowledge in the simplest, most linear manner—in other words, via a revisionist description that differs significantly from the wandering path that scientists actually followed. This is not misguided—to complicate further the already formidable challenge of clearly transmitting scientific facts would be counterproductive. But there is a resulting trade-off: After years of absorbing scientific facts taught in this way, it can be a considerable challenge to learn how to practice the art of scientific research, which is, in actuality, much messier.

The reality of Jacob's night science is that its practice is much more iterative, uncertain, and fitful, but at the same time much more exciting and personally satisfying. It really is a creative process, full of struggle punctuated by inspiration. Like all creative processes, design is involved at all steps. Take the ostensibly straightforward, logical formulation of the "simplest hypothesis that can explain the data." How do scientists formulate (or design) the simplest hypothesis? Often, no single hypothesis is clearly the simplest, so how are these hypotheses broken up into families with common core elements? How do scientists figure out which ones are most easily testable or which ones, if disproved, would be most informative in limiting other possible models? It is not as straightforward as it would seem, but grappling with these questions is incredibly rewarding.

Science as a craft is at once both enormously creative and cautious. The importance of producing sound, evidence-based conclusions leads many within the scientific community to view speculation and conjecture with suspicion. Yet in order to derive these sound, testable conclusions, scientists need to know what questions to ask and what experiments to do next; for those purposes, speculation is essential. Night science is extremely difficult to practice and to teach—examining it as a creative process could be very fruitful. But instead of thinking more critically about night science—dissecting it and breaking it down analytically as designers have done with their practice—there is a tendency to suppress it and replace it with a sanitized day-science caricature.

Insights on Science and Design

Design does not exist in a world free of constraints either. Any successful design process implies an outcome, a deadline to meet, and usually a client to satisfy. In this way, science and design are actually rather similar; both disciplines seek to define visions that challenge the status quo, but are constrained in terms of how these visions can be conveyed. But though these constraints apply to the final object (a scientific publication, a designed object), they need not apply as stringently at all points of the process. In our conceptual deep dive, both designers and scientists were afforded unusual freedom from the constraints of their everyday practice.

Many design projects begin with a creative stream of consciousness, in which all ideas are given a chance to flourish and be free from judgment, resulting in a barrage of apparently random and abstract concepts. A series of prototypes will often be produced relatively early in the process, only to be discarded further down the line. At the conclusion of our deep dive, participants presented their ideas to the group, though they were not fully fleshed out. This experience was challenging for the scientists, who were somewhat hesitant to share ideas that they had not fully processed.

Designers at IDEO have put tremendous effort into articulating and modeling their creative process in as much detail as possible. This is because design, not unlike night science, is inherently a loose and constantly changing journey on which one can easily become lost. By adding structure to this journey, designers shape it into a controlled and manageable process and are able to move back and forth between exploratory and more focused ways of working. Any design project can move through multiple cycles of this process. The diagram of our design process shows the journey in a way that resembles a funnel: as a contorted pipeline with cycles of expansion and contraction, moving from insights to synthesis, and from opportunities to concrete ideas (figure 9.2).

Nearly every client goes through stages of frustration with the unfamiliar process of design. When what seemed like a fairly targeted initial idea has morphed into what appears to be a diffuse grab-bag of abstract concepts, clients typically begin to wonder if the seemingly directionless jumble that has emerged is worth the fee they are paying.

One of the most interesting points consistently highlighted by designers is that periods of great uncertainty and frustration are an inevitable part of the design process. Designers know that the design process will involve cycles of idea expansion and contraction, and they actively manage the flow of the project to encourage this.

When Will and Adam—the IDEO designers—first drew this "design funnel" for the Lim lab scientists, it became apparent that this depiction could

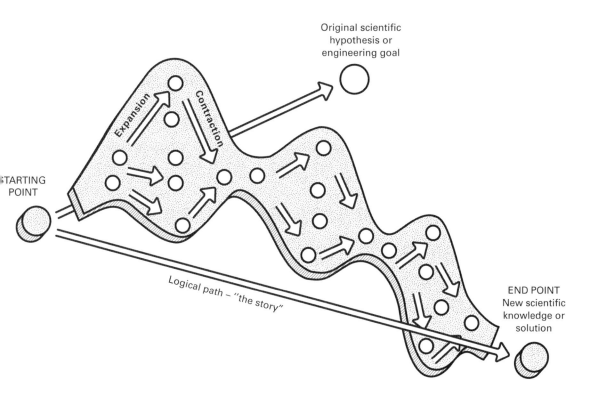

Original scientific
hypothesis or
engineering goal

Expansion

Contraction

STARTING
POINT

Logical path – "the story"

END POINT
New scientific
knowledge or
solution

just as easily describe the progress of a scientific project. It gave shape to the reality of night science in that scientific projects are constantly being reshaped and redirected and their underlying hypotheses expanded, changed, and contracted by the models considered. Uri Alon has referred to the frustration and uncertainty that comes with these cycles of change as "being in the cloud."[3] The cloud is something that inexperienced scientists fear and can cause them to become paralyzed. But for the experienced scientist, it is an expected stage that, if you keep your wits about you, often presages remarkable advances. The hidden story of the successful scientific process can be represented by a very similar "digestive tract" diagram, full of exploratory idea expansions and critical contracting cycles driven by empirical testing (figure 9.7).

An important outcome of this project is the dawning realization that scientists are not doing themselves or their students any favors by suppressing the struggle inherent in the scientific process. Ironically, when scientists do discuss the vicissitudes of night science, there is a tendency to mythologize it—to describe it in a mystical and surprisingly nonscientific way. If night science can be brought out into the light and examined in a more open and dispassionate way—if it can be dissected into its general and expected steps—young scientists could be provided with the tools to work through the confusion and frustration it inevitably entails. Science will always require a

Figure 9.7
Wendell Lim's reinterpretation of the IDEO design process to describe the scientific process.

combination of many elements—a broad knowledge of prior established facts, facility with difficult techniques, and cold, logical analysis—but perhaps a little more thinking about the process of scientific creativity and design could have a major impact on scientific practice and training.

Collaborative Commitments

Our aspiration to "design nature" was as audacious as it was experimental, and perhaps too great an expectation for a brief collaboration such as this. However, the potential for interdisciplinary cross-fertilization, network building, and mutual learning that was unlocked through our interaction represents a legacy that is still unfolding.

Our collaboration has ignited an ongoing series of high-level discussions at both IDEO and UCSF concerning the prospect of further opportunities for synergy between science and design, and we very much hope that other institutions and individuals will follow in our footsteps. Our intention has always been to provide a more colorful window onto the complex subject of synthetic biology, usually confined to academic journals, illustrating its potential in a more tangible and accessible way to enable more public debate. We have therefore sought to publicize the results of our collaboration through both scientific and design networks.[4]

For the designers, this project represented an opportunity to explore the unfamiliar territory of synthetic biology and to imagine the possibilities that such technology permits. We hope, through the various publications stemming from this collaboration, to inspire other designers to engage their potential not just in using pioneering new technologies but as active participants in both the development of said technologies and the debate surrounding their economic, social, and ethical implications. Meanwhile, the project has offered the synthetic biologists involved a space in which to envision new possibilities in a visceral way that day-to-day laboratory research does not often permit. Interacting with the designers also provoked the scientists to reflect at length on their own methods of working and to imagine new ways of teaching and inspiring the scientists of the future.

10.
Constrained Creativity: An Engineer's Perspective

Alistair Elfick

It is easy to imagine the myriad existing products that may be readily displaced by surrogates produced from synthetic biology: liquid fuels for transport shall be derived from biomass rather than refined from fossilized biology; drugs will be synthesized in bulk by microbes instead of extracted in tiny quantities from plants. Such "drop-in" replacement products will be adopted, unnoticed, by the consumer, save for the manufacturer's assertions of the sustainability of its product.

But what of the unimagined uses of synthetic biology that could emerge? This was one of the questions that motivated the interaction of the *Living among Living Things* team, designers Will and Adam and scientists

Wendell and Reid. They postulate that through speculation of possible futures, we may negotiate the degree to which the uses of a technology may be allowed or indeed disallowed. The assertion that through picturing those contemporary futures we stimulate discussion about the uses of synthetic biology is one that has been accepted by sections of the synthetic biology community. This process will also bring a richer realization of that which synthetic biology actually delivers. There is a willingness to believe that the promise of synthetic biology can deliver, must deliver, far more than the discrete displacement of unsustainable manufacturing practices.

In chapter 9, the team members recount their exploration of synthetic biology and design. The most readily identified early challenge was that of communication: the need to arrive at a common rubric and appreciation of each other's ways to give a footing for the collaboration to begin. An exploration of communication between disciplines is given by Pablo in chapter 8.

Living among Living Things is predicated on the notion of implementing biological products that exploit mutually beneficial relationships with humans. This concept may seem radical at first, yet our history is littered with symbiotic interactions; anyone owning a dog has entered into such a contract with a biological consumer product, though he or she may not recognize this. What leaped out of the team's thought experiments was not the learning to be found in the future concepts produced, but rather the unexpected similarities between the descriptions of the design process and the actual practice of science—Jacob's "night science."[1] The description of the cycles of uncertainty and change described by Alon's "cloud"[2] seems to apply equally well to design or science, or for that matter to any creative endeavor. Characterizing science as a creative process may jar with the notion of rigor and a linear logical process, yet what they refer to as the "art of scientific research" surely exemplifies skill, celebrates quality, and generates value in just the same way as the "authentic craftsmanship" in design. The similarities in process are striking; the variable seems to be the output generated—either artifact or knowledge. What leaps off the page is the sense of pride in their professions that both sides display and how personally rewarding they find their jobs.

The "cloud" can be an uncomfortable place for scientists; frustration and uncertainty are difficult to manage even for the experienced. For the inexperienced scientist, this can easily be "paralyzing," as asserted by the team. The success of IDEO as a design consultancy is based on an acceptance of the confusion of the cloud. By taking the team through a "design deep dive," the scientists were given a first-hand experience of managing the process of conceptualization. The cyclical process of expanding an idea to a "grab bag of abstract concepts," then contracting toward a solution before expanding again, maps well onto Jacob's night science. The key would appear to be having the confidence to persist with this process. Embodying this cyclical nature as "part of the process" gives a sense of security, encouraging the participants to embrace the freedom of the cloud. Neither design nor science is free from constraint (client demands, cost, time, funding): Making space to swim free in an ocean of possibilities must be treasured.

Jacob's "day science," the reconstitution of the scientific process, its sanitization for ease of dissemination, yields clear communication but may also mask subtlety, even among the informed peer group for which it is intended. It could be argued that this process of repackaging is not constrained to sciences. In reacting to a work of art, the lay public is insulated from the years of learning, the hours of concentration, and the raw physical energy that was invested in its creation. The reaction of a fellow artist to the same artwork will necessarily be different, tempered by his or her knowledge of the process; the appreciation transits from aesthetic to additionally technical. When using the latest electronic gadget, a consumer seldom ponders the thousands of person-hours devoted to every aspect of its production and distribution: these are hidden. These seem very disparate notions, but at their core is the notion of appropriate facilitation—enabling understanding of scientific facts for scientists, affording the viewer freedom to react personally to a work of art, allowing a designer to appreciate technology in a product, or giving ease of use to the consumer.

The distancing of an audience from the process serves to deprive that audience of a fully informed response; this may be inadvertent or intentional and can have positive or negative consequences. In any case, it disenfranchises; only those learned in the process of science/design/art can experience a full aesthetic or technical appreciation, and hence a full right to critique. What emerges from the work of the team is the realization of the ramifications of this practice in science. They identify difficulties caused by the transition of students from the didactic training practiced from kindergarten to undergraduate study to the cloudy world of research. But the consequences of the disconnectedness implicit in disenfranchisement may also serve to frame the way that the lay public experiences synthetic biology. If denied the opportunity to arrive at an informed opinion, there is the likelihood that views will lack the nuance necessary to negotiate how synthetic biology may be deployed. There is the prospect that public opinion becomes polarized and synthetic biology viewed wholesale as a good or bad thing.

Capturing Creativity

What of those products of synthetic biology that are at risk of being inadvertently precluded? There exists the possibility that through embracing the cloud and fostering creativity, the truly transformative innovations of synthetic biology will be discovered. We must ensure that we don't unwittingly foreclose the possibility of some potential future in the way we institute and regulate synthetic biology. A simple thesis presents itself: If we fail to preserve space for creativity within synthetic biology, we will be impoverished.

Creativity is a concept that, at face value, appears simple. However, in practice creativity takes a myriad of forms; the creative output may be tangible or esoteric, can be artifact, process, or idea, and should include both the useful and the useless. To discuss the role of creativity in synthetic biology effectively, it is useful for me to adopt a working definition of the term, derived from *Webster's* unabridged dictionary: "Creativity is the endeavor to transcend traditional ideas, rules, patterns, and relationships, and through embracing adventure, generate meaningful new ideas, forms, methods, and interpretations."

This definition of creativity does not attempt to discriminate it from innovation, which could be considered to be an instance of creativity focused on functional advancement. Creativity would seem to be a peculiarly human trait. Over thousands of millennia, humankind has demonstrated a propensity for creativity, in everything from tool use and cave art to personal computing and cosmetic surgery. The creative impetus would seem to be innate and diverse, but without cultivation it withers.

Every parent marvels at the creativity of his or her child, be it in imaginative role-playing games, drawing, or story-telling. The willingness to explore and the lack of restraint that can be found in naivety are liberating. Yet something happens on the journey from child to adult, which for many results in the loss of their creative curiosity. There are doubtless many influences that conspire to affect this outcome, but one that may be readily identified is the educational system. In 1959, C.P. Snow[3] presented an essay that identified "two cultures" in British education—the humanities and the sciences. He argued that at an early age, there was a compulsion to choose between these cultures, subdividing the arts from the sciences, the creative from the reasoned. This enforcement of disciplinary silos is still something that challenges educational systems, and perhaps the British one more than others. It is evident that this will act to reinforce the perceived disconnect between science and creativity.

Plato viewed creativity as a form of divine inspiration, a kind of madness.[4] This has become a widely held perception reflected in popular culture. We are all party to the notions of the odd-ball painter, the mad scientist, the eccentric inventor, and so on. The French mathematician Henri Poincaré gave a compelling account of his own experience of the creative process.[5] He describes ideas arising in swarms and combining seemingly randomly in his subconscious with him selecting the most promising ones according to "aesthetic criteria." Psychological studies have shown a correlation between irrational behavior and creativity. Irrationality may lend itself to the breaking of boundaries, revealing rich sources of new ideas. It is possible that such

DESIGN FOR BASCULE BRIDGE OVER THE THAMES.
[BELOW LONDON BRIDGE.]
II. BRIDGE CLOSED.
By HORACE JONES, CITY ARCHITECT
PRESIDENT of the ROYAL INSTITUTE of BRITISH ARCHITECTS

overinclusive thinking breaks down distinctions and generates the kind of free association that propels creativity. The selection of promising ideas based on some sort of "fitness landscape" has led to the proposal of a Darwinian notion of creativity (figures 10.1 and 10.2).

The dominant paradigm within synthetic biology asserts that there are clear benefits to bringing a formalized engineering process to the way we endeavor to make with biology. In doing so, there is the tacit assertion that modern engineering practice is at a zenith. Yet the role of creativity in twenty-first century engineering is far from clear. What opportunities for creative expression are available to the individual engineer in a large company?

To address this question, we need to reflect briefly on the history of the profession of engineer and the role of creativity. The "practical art" of the engineer's antecedents embodied imaginative tinkering and development through trial and error. Over time, engineering evolved and metamorphosed, from a craft-like pursuit into a profession as the result of the Scientific

Figure 10.1
Natural selection is the key mechanism of evolution.
A parallel can be drawn to the process of selecting the fittest idea for a design, or rather by rejecting those designs believed not to be optimized. For example, the Tower Bridge in London might have looked very different.
It took 8 years to decide on the chosen design, an evolution of Horace Jones's design seen here.

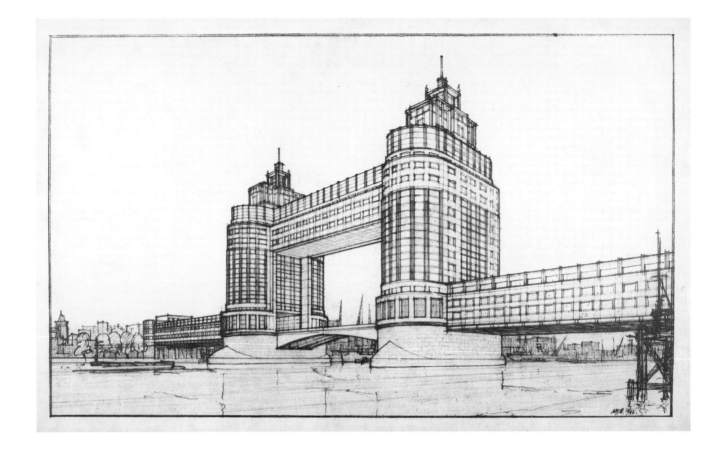

Figure 10.2
Shown is a 1943 design proposal
for a glass-enclosed retrofit of
Tower Bridge by W.F.C. Holden,
this in case of wartime bombing.

and Industrial Revolutions of the sixteenth and seventeenth centuries and eighteenth and nineteenth centuries, respectively. The earliest adoption of formal training in engineering was within military academies; the French *corps du génie* founded in 1676 facilitated the accumulation of technical knowledge, skill, and practice within a distinct grouping. The precursor of the professional engineer was the officer class of such military bodies. Civil engineering emerged as the civilian deployment of military technology. Over time, apprenticeship was replaced by academic study, with degrees accredited by professional institutions. Modern engineering is a highly technical discipline founded in scientific discovery. Across all engineering disciplines, there underlies a common tenet; that engineering is, in essence, a communal endeavor.[6] The implementation of standardization, and abstraction of function, was inherited from the military along with the adoption of a hierarchical working structure (see figure 1.6 in chapter 1). Specific roles within an engineering task were given to particular sets of individuals. The subdivision of responsibilities becomes increasingly attractive as projects become more and more complicated. The adoption of specific roles created the opportunity for design to become gradually separated from implementation. In the early twentieth century, industrial design emerged as an approach to understand the influence of a product's form and function in mediating interaction with the user.

How does creativity sit within this framework of engineering? In the separation of engineering and design into discrete practices, the opportunity for creativity was not equally shared. Engineers have seen their connection with creativity eroded, their ability to invent and innovate constrained. It is my contention that engineers are impoverished; creativity and self-expression have been too often subcontracted to the designer, architect, marketing consultant, and so forth. Invention has been relegated to be the domain of the eccentric amateur. The profession of engineering has turned its back on tinkering and play; bereft of creativity, we have become the implementers of the inspiration belonging to others.

If engineers have been impoverished, does it hold that designers are enriched by the ring-fencing of creative expression? Not necessarily. Designers are often constrained by lack of full engagement with technology. The link between form and function has become weakened. A change in emphasis occurs; design becomes pushed by technology rather than pulled by functional requirement, and technology loses value through weak design input. The need to monetize technology takes precedence in the design process; a "hi-tech is good" mindset prevails in the consumer. Design is being co-opted for the realization of desirability, and away from the pursuit of eloquent function. Creativity has been harnessed to provide a particular role in commerce and society. It may be asserted that creativity is the fuel that drives the engine of capitalism. Products are perpetually invented and reinvented; obsolescence is produced, need is generated, desire is fashioned. There has been a change in emphasis: from the use of technology as a way to fulfill a need, to the creation of need using technology. Product development is often driven not by design obsolescence but by the commercial expediency of having something new to market. Design risks becoming the slave of product identity with function sharing the billing with fashion and the cult of brand.

The field of synthetic biology embodies the transition from designing and making using the materials of nature, as we have for millennia, to the design of biology itself. Genetic modification (GM) of the late twentieth century was a practical art, highly skilled, artisan, and craft-like. DNA was tailored for bespoke production of high-value products. There is then clear benefit to be gained from capturing the craft of genetic modification, its learning and infrastructure, and sharing this with a community of biological engineers. In rethinking twentieth century biotechnology, and in the application of the principles of engineering and design practice, there lies the unprecedented opportunity to facilitate a mode of practice that seeks to preserve the joy of play, the wonder of discovery, and the thrill of creativity.

How can we ensure that synthetic biology is instituted in such a way as to embrace creativity? In framing the *engineering* approach to genetic modification, analogies to the Industrial Revolution are used to exemplify the power of deskilling manufacture. Standardization and abstraction enabled the encapsulation of tasks and information, propelling the subsequent move to mass manufacture and mechanization. Yet the inherent complexity of nature puts us on a back foot. We are not starting with a simple material and building complexity; quite the opposite is true. The fabulous wilfulness and malleability of biology presents challenges that engineers have never before faced, such as the evolution and reversion of function. Aluminum seldom spontaneously develops the ability to flow at room temperature, nor decides it would rather be ore again.

In defining synthetic biology in chapter 1, we reflect that it should not be considered to be the deskilling biotechnology but rather the "up-skilling." The craft of biotechnology can be captured in a framework that allows the pooling of knowledge and its gifting to a community of biological engineers. By capturing a base level of knowledge and craft, synthetic biology aims to deliver a step-change in the ambition of our endeavors, but this will not, in itself, foster creativity. Up-skilling will deliver the more efficient and rapid realization of our known ambitions, those drop-in products identified earlier as the low-hanging fruit of sustainable manufacture, but the next stride, the unanticipated futures, will not necessarily follow. This requires the stimulation of creativity within our systems of work. Synthetic biology must engender a situation where creativity and innovation, discovery and play, utility and uselessness are encouraged to coexist flourishingly.

As an engineer, one is trained to apply logic and rationality in pursuit of a particular goal. There is little scope to indulge fancy or frivolity as practicality is our currency. This approach has been characterized as "parallel thought":[7] logically based, with conformity and compliance as its bedfellows, giving a sense of certainty and the illusion of knowledge. The antithesis of parallel thought is "lateral thought": daydreaming, fantasy, thought free from the constraints implied by implementation but also free from the security of precedence.[8] These divergent thought processes are difficult to reconcile, but there exists great potential in playing them off each other. Derailing parallel thought with judiciously applied acts of provocation through lateral thought is an oft-employed tool to stimulate creativity. This is akin to the IDEO process (depicted in figure 9.2 of chapter 9), which was adopted in the placement. This hybrid mode may be considered "step-wise thinking," an effort to temper the stimulation of thinking free from prior assumption with a focus on taking the fruits of these thoughts to fruition.

To be effective, lateral thought requires a certain environment. The practitioner must have a competence within the domain in which he or she wishes to ideate. Free-thinking exploration should be driven by curiosity and be rewarding in and of itself. The freedom to explore should not be limited by a reticence to fail; risk must be embraced, even encouraged. The opportunity for choice and discovery necessarily creates the possibility of failure. However, it should be recognized that a tension exists between freedom and constraint, both of which can equally stimulate or retard creativity; for example, funding constraints can simultaneously promote inventiveness and limit opportunity. Indeed, the very nature of community-based engineering, with its standards and protocols, legislates against the inclusion of lateral thinking and creativity.

Failing Confidently

It is widely recognized that the possibility exists for lock-in to particular technologies. A great example of this is the layout of the QWERTY keyboard, which was devised to reduce typing speed in order to limit the possibility of jamming in early mechanical typewriters.[9] While technology has moved away from typewriters to personal computing, we are locked in to the QWERTY keyboard format. Promoting creativity and innovation in synthetic biology may require occasional "molting." Just as crustaceans need to shed their exoskeleton to grow, synthetic biology must not be reticent to undergo a step renewal of our technological underpinnings. Some of the most creative periods in the history of art have centered on the deconstruction of dogma, such as the emergence of abstract art and its challenge of the representational. In generating operating standards for synthetic biology, we must accept that they may be necessarily transient. We must be ready to throw away our successes alongside our failures.

There are companies that have shown that investing in play and intellectual freedom can stimulate innovation: The IT company Google gives its employees 20% "free" time to foster creativity. Synthetic biology can be enriched by preserving the sense of fun that it has captured to date; embedding curious tinkering, while preserving the cornerstones of safety and responsibility, will maximize the opportunity for the guided serendipity from which great discovery has often come.

The potential of playfulness in synthetic biology is exemplified by the activities of the participants in the International Genetically Engineered Machine (iGEM) competition. Since 2005, teams of undergraduate students from across the globe have been assembling to display their achievements after a summer of designing, building, and testing a piece of synthetic biology. The diversity of projects is now routinely breathtaking. Underlying all

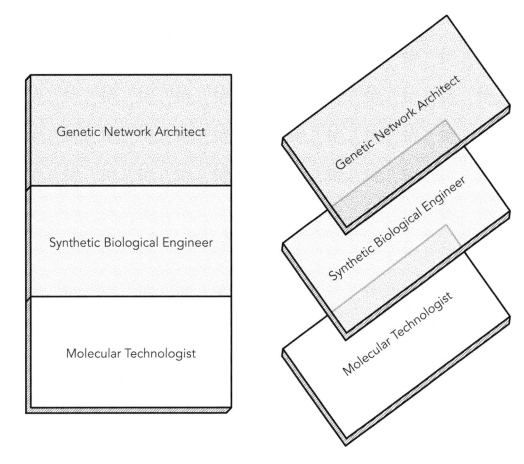

Figure 10.3
Blurring the boundary between
one level of the abstraction
hierarchy to the next could serve
to maintain more freedom for the
practitioners of synthetic biology
to innovate.

these projects is a sense of optimistic naivety, which is seldom encountered in professional research laboratories. The possibility for discovery and creativity inspired by play and tinkering exists at present, but there is a risk that the door will close. Only if the cost overhead of play is minimal, and paid back with interest through beneficial output, will curiosity-driven tinkering be maintained once synthetic biology becomes embedded in commerce. Can synthetic biology be cheap enough to play? Probably. The reducing cost of DNA synthesis is already enabling a depth of ambition not seen previously; synthesis will be inexpensive enough to enable tinkering and play by removing the cost of nonproductivity.

It is clear that the framework of engineering used since the Industrial Revolution has been highly successful. However, the engineering of nature is quite a different proposition, and there is merit in challenging the assumptions of the abstraction hierarchy (see figure 1.6 in chapter 1) with its tacit introduction of technical specialism: designer–engineer–technician. Difficulties become manifest with the creation of silos of expertise with paucity of communication across their boundaries, thus limiting what may be achieved. In instituting synthetic biology, there is surely merit in considering a more "tilted hierarchy" of skills; one in which each practitioner must achieve a more vertically "stretched" set of expertise. Preserving overlap in knowledge and capability at the interface between practitioners will act to reinforce communication and create space for creativity and playfulness.

In aspiring to embed creativity within synthetic biology, we propose a manifesto for the liberation of engineers that not only seeks to enrich our subject but also has the potential to impact all of twenty-first century making (figure 10.3).

Time

11.
The Biogenic Timestamp:
Exploring the Rearrangement of Matter through Synthetic Biology and Art

Oron Catts and Hideo Iwasaki

In the light of increasing human intervention and influence on geological and biological processes, there is a need to question the underlying hubris of human intentions to control life. In this context, we hope to demonstrate that time can be used as an instrument for humility. As a way to examine the use of time, our research involves exploring different ways of manipulating the cyanobacterial biological clock, its spatiotemporal pattern formations, and its ability to accumulate, deposit, and precipitate metals and other substances. Cyanobacteria engage with time in many different ways, from the rhythms and buildup of events and movements ranging from rapid microscopic actions, to day–night cycles, to extremely slow macro scale/geological formation. The

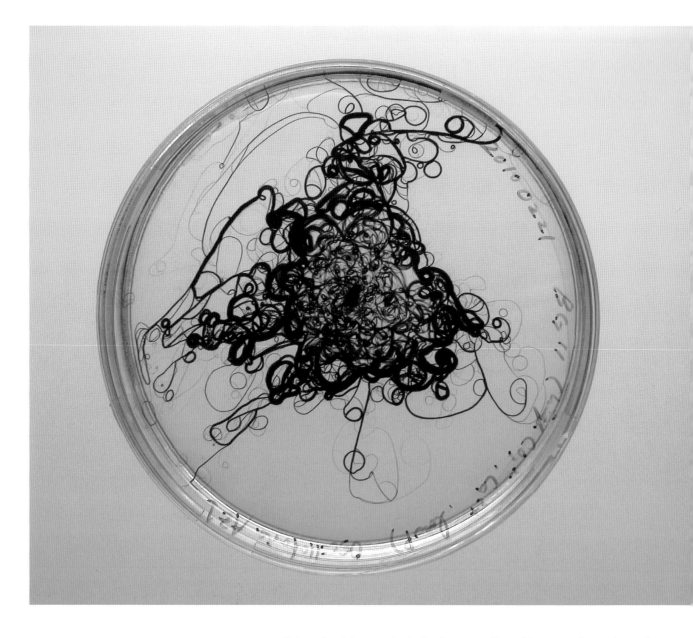

concept of time (and its manipulation) as manifested by cyanobacteria and their by-products represents a fertile ground for the exploration of different modes of artistic practices (figure 11.1).

Much of Earth's crust is biogenic, meaning it was produced in association with living organisms, mainly cyanobacteria and their relatives. For the greater part of the existence of life on Earth, cyanobacteria have been the dominant life form, transforming Earth's crust and atmosphere and leaving traces of their activity as sedimentary layers of rock. Now humans

Figure 11.1
A colony of the filamentous cyanobacterium *Oscillatoria* sp. that Hideo works with.

are doing much of the shifting and depositing of materials. The changes we make in Earth's crust are linked to changes in the planet's atmosphere, and we are already having major effects on life and climate. This activity is being posited as a new geological age, the Age of the Anthropocene; a time in which anthropogenic (human-related) formation is becoming the dominant feature of geological, geochemical, and biotic shifts. The Anthropocene ultimately represents the effects of human industrial activity, "encompassing novel biotic, sedimentary, and geochemical change."[1] Higher levels of atmospheric CO_2 are leading to acidification of the oceans and a dissolving of previously laid biogenic carbonate sediment,[2] erasing geological evidence of earlier biogenic geological formation. However, in the near future, through developments in synthetic biology, we may well witness other novel forms of biogenic sedimentation.

The complexity of the cascading temporal and evolutionary events resulting from what we consider now as premeditated human actions will in all likelihood be seen, by far future reflections, as a time in which random mutagenic and geological agents roamed free. In a few million years, the human era may be just another thin layer in the rock formation, a humbling thought.

Will the unintentional outcomes of human action be not too different from the effects of the "discovery of oxygenic photosynthesis" by cyanobacteria? Photosynthesis allowed life to be autotrophic (i.e., self-sustainable) and even excessive: to create a surplus of nutrients that supported other life forms. It is claimed that photosynthesis had a major effect on the transformation of Earth's atmosphere into an oxygen-rich one, poisoning many of the anaerobic life forms, but in turn, allowing the evolution of fast-metabolizing aerobic organisms.

On a poetic and metaphorical level, the comparison between cyanobacteria's autotrophic existence through the "discovery of photosynthesis" and human technology could become a site for contemporary artistic intervention and cultural scrutiny. It would be interesting to run a parallel and comparative artistic analysis of the "discovery of photosynthesis" alongside synthetic biology's attempts to transform cyanobacteria into a piece of "rationally designed" technology, mainly for biofuel and hydrogen production.

As previous chapters have explained, synthetic biology is an emerging interdisciplinary field of biology/biotechnology that aims to "design" new biological functions ranging from in vitro reconstitution of cellular functions to extended genetic engineering to modify organisms and even to make synthetic cells. Above all, it aims to make "useful" biological tools, to understand current life forms, and to explore the origin of life. It also raises philosophical questions about the nature of life and the boundary between life and inanimate materials, as well as societal questions about biosecurity and bioethical

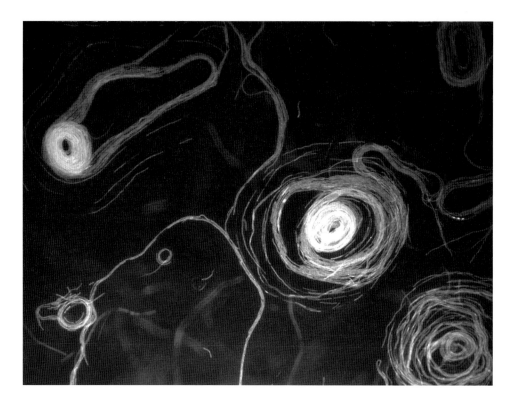

issues. Thus, it is a new technological field in which artists and designers can play important roles in considering contestable futures and exploring the cultural and societal implications in ways that differ from scientific and industrial viewpoints. Synthetic biology has the potential to change organisms without historical precedent, and the cultural significance of its use for ecological and geo-biological studies has been largely ignored.

Our artistic approach attempts to "engineer futility" into cyanobacteria, by interfering with processes and "efficiency," creating a kind of thoughtful absurdity meditation over the idea of time as an instrument of humility. Of particular interest to us here are circadian cycles, photosynthesis, and the ability of cyanobacteria to precipitate different substances.

Our research focuses on the links between geological and biological time in relation to the logic that drives synthetic biology. We explore different aspects of the material manifestations of temporality and placement using an array of performative artistic practices and new developments in synthetic biology, working with cyanobacteria as model organisms (figure 11.2).

The earliest fossil record on Earth is of cyanobacteria (formerly known as blue-green algae). As mentioned earlier, cyanobacteria are the first autotrophic life forms and were responsible for making Earth's atmosphere oxygen rich and also for much of the sedimentary rock formation. Cyanobacteria are

Figure 11.2
Autofluorescent (photosynthetic) image of the filamentous cyanobacterium *Geitlerinema* sp. that Hideo researches.

increasingly becoming an organism of interest in research and for application as they seem to have potential to be used for biofuels, hydrogen production, food production, environmental control, and metal accumulation.

All biological processes are time-dependent across a variety of timescales. Cyanobacteria are especially interesting from this point of view because their photosynthetic activity is based on extremely rapid physicochemical electron transfer within an order of picoseconds (one trillionth of a second). Their cell cycle ranges from 10 hours to more than 10 years, depending on species and environmental conditions. Their internal circadian clock precisely times and gates gene expression and metabolic activities over a 24-hour period. Cyanobacteria populations also accumulate calcium carbonate (and possibly other minerals) to make sedimentary rocks (called stromatolites or thrombolites) and precipitate metals over a much longer geological timescale, which can stretch to thousands of years (figure 11.3). The development of synthetic

Figure 11.3
Thrombolites in Lake Clifton, Western Australia. These "living rocks" are produced by cyanobacteria over hundreds of thousands of years.

biology provides a situation in which some of the oldest living organisms are to be engineered and manipulated for human ends, merging the biogenic with the anthropogenic.

Western Australia is a unique place in that it is a meeting point between the biogenic past and future. The Pilbara region in Western Australia holds one of the oldest rock formations on Earth, known as the "Pilbara craton," which is one of only two pristine ancient geological formations in the world (formed 3.5 billion to 2.7 billion years ago). This region offers one of the best-preserved "geological calendars," at the Hamersley Range. This is also the region where some of the largest deposits of iron ore are currently being extracted (figure 11.4), in one of the fastest and biggest biologically induced re-formations of Earth's crusts. Pilbara is now also home to one of the biggest commercial algae production sites in the world, Aurora Algae in Karratha, where "genetically optimized" algae are grown to produce pharmaceutical products (omega-3 oils), food, and biofuels.[3] Western Australia

is an exceptional region for its living microbiolites, such as stromatolites and thrombolites, which are still active in some lakes and pools. However, as in the case of Lake Clifton (figure 11.3), the thrombolites are under threat due to climate change and urban development.

Our project involves artistic research in some of the places of biological and biogenic geological significance, as well as lab work to explore the possibility of human-induced biogenic formation. By considering geological and human-derived biogenic formations, our artistic research provides an alternative use and critical interpretation of synthetic biology. Attempting to use the abilities that some cyanobacteria might have in accumulating and assembling nanoparticles of gold, iron pyrite (also known as fool's gold; figure 11.5), and silica, we are developing a range of artistic practices. The best of these will be further developed and incorporated in situ (rock formations, labs) and in artistic contexts (galleries, festivals, publications). An example of one of the research trajectories involves manipulating and culturing cyanobacteria on computer motherboards, which was exhibited for the first time in October 2012 (figure 11.6). Our aim is that the organisms will rearrange and deposit some of the components, such as silica, gold, and iron, and in this way present a symbolic case in which life "messes up" the linear logic of digitally driven synthetic biology.

Figure 11.5
Iron pyrite, also known as "fool's gold."

Figure 11.6
Installation view of Hideo's
cyanobacteria dissolving
printed computer boards in the
BioAesthetics exhibition in Tokyo,
October 2012.

Synthetic biology promises fundamentally and radically to change the way we relate to and treat life—conceptually, materially, and temporally. The question of life is one that has long occupied artists, so it is not surprising that artists are drawn to synthetic biology (regardless of whether its promises are fulfilled or not). Another promise of synthetic biology is to deliver tools that will make it easier to manipulate life—through the engineering logic of standardization, deskilling, and the idea of the "black box." This means that artists can start working with life as a raw material for their projects, having a new palette of possibilities of manipulating matter, in this case, life itself. In other words, "life" has always been a matter of art. If we consider mimicking organisms as gestures toward inventing artificial life, then the first drawn animal/plant images may be interpreted as the origin of some aspects of synthetic biology. Thus, our research is a ritual to connect the earliest and contemporary attempts to reformat the historical timescale of our relationship to biological existence.

As far as this project deals with the idea of time of different scales and the idea of time as an instrument for humility, it helps to be challenged by Manuel DeLanda to "Imagine an observer with a time-scale so large...He wouldn't even see us. Species to him would seem like vast amounts of biomass in constant change. That observer would see species mutating and flowing. He would probably worship flows—unlike us, who, because of our very, very tiny time-scale of observation, tend to worship rocks."[4] Furthermore, a single drop of water from ponds, lakes, rivers, or oceans contains a variety of species of cyanobacteria as survivors and recorders of the whole of biological/geological evolution.

Finally, we would like to refer to a strange coincidence in the topics addressed in the Biogenic Timestamp project with some problems revealed by the disastrous tsunami and the subsequent nuclear power plant incident in Fukushima, Japan. This case demonstrates the importance of "time" ranging from nuclear reactions at the quantum level, to long-lasting disposal of radioactive waste from power plants, to mutual interactions between industrial development, enhancement of social activity, usage of natural resources, and human desire, as well as the human/nature-induced dramatic environmental change, and the effects on daily life at different timescales.

The authors have a long-standing interest in combining biology and art through their respective disciplines; working with different levels of biological processes and the broader cultural implications of such activities.

Hideo Iwasaki has been working on the mechanisms underlying circadian timekeeping and morphological development in cyanobacteria. In circadian studies, Hideo has reported identification of "clock genes" (core parts of the body clock) and recently succeeded in making an "oscillating soup" with a period of 24 hours by mixing the core clock proteins in a test tube. In vitro reconstitution of cellular functions has been attempted using genetic information, biosynthesis, and cellular membranes. By 2004, it became clear that synthetic biology would proceed at least in part toward a synthetic cell. Hideo and colleagues founded a domestic society for cell synthesis research in 2005, in which he has served as an organizer of a unit of philosophical and societal implications of the field. From 2007, Hideo founded an in-lab biological/biomedia art platform (later named metaPhorest), in which some artists share benches and experimental facilities with scientists in his lab.

An extension of this chronobiological study is aesthetic/scientific research on spatiotemporal pattern dynamics in cyanobacteria. For example, Hideo has attempted to construct three-dimensional architecture/sculpture using cyanobacterial culture isolated from his neighborhood in Tokyo and has shown that a film of filamentous strain, *Nostoc* sp. is occasionally able to form a bubble structure via self-generative oxygen evolution. Hideo then improved a protocol for more reproducibly inducing cyanobacterial bubbles on plates in an artistic project, *CyanoBonsai* (2009). The bubble structure formed is relatively stable, lasting for more than a month on agar plates, and probably involves a biosedimentary process. Notably, another type of bubble formation in another cyanobacterial species has been suggested as a possible initial step toward stromatolite formation.

Oron Catts has worked as an artist and researcher within the interface of wet biology and culture for the past 16 years. He has a keen interest in critiquing the ways in which development in the life sciences has been "sold" to the public. Synthetic biology is a contemporary example of a field that employs artists and designers as part of a concerted effort to engineer public acceptance for a technology that does not yet exist. He identified synthetic biology as an area that requires particular cultural scrutiny.

His previous work dealt mainly with the shifting relations to life through the engineered fragments of living bodies. Working since 1996 under the banner of The Tissue Culture & Art Project with Ionat Zurr, they have explored both symbolic and pseudo-utilitarian projects such as growing in vitro meat, leather, and pig wings. Since 2008, Oron was also involved, in his capacity as the director of SymbioticA and as an artist, with a long-term research project investigating one of the last remaining active thrombolites sites—Lake Clifton in Western Australia.

One of Oron's artistic projects, *The Autotroph*, is a solar-powered kinetic sculpture and fountain, inspired by Lake Clifton (figure 11.3). As mentioned earlier, the lake and its inhabitants, including the thrombolites, are under threat due to salinity, urban development pressures, global warming, and land misuse. *The Autotroph* consists of a floating solar still; it was exhibited floating in natural and artificial bodies of water, using the rays of real and artificial suns. The still deployed a computer-controlled focusing mirror to achieve sufficient heat to evaporate water. The steam generated was condensed and tipped back into the body of water. It is an overly technological and playful exploration of the immense complexity of dealing with ecological issues. It ironically explores the problems and possibilities of technological solutions to human-induced climate change. Any action postulated raises possibilities of good and harm to different aspects of the ecology, and this is without even considering the unknown unknowns (to quote Donald Rumsfeld). The challenge of this project is to tell the stories of these complexities, but not to solve them, as is the case for the Biogenic Timestamp project.

12.
Time as Critique

Jane Calvert

Reports on synthetic biology often speculate about its future, typically attempting to predict what applications we might see in 5, 10, or perhaps even 50 years' time. But what happens when we think about how synthetic biology might look from millions of years in the future? Will we find a layer of rock with fossils of synthetic biological creatures? Will the living legacies of synthetic biology be roaming the landscape? Or will synthetic biology be forgotten and unimportant for a future archeologist?

It is questions like these that are raised by the microbiologist Hideo Iwasaki and the bio artist Oron Catts in their collaborative work. By challenging us to think about synthetic biology from the perspective of geological time—the timescale of geological events since the formation of Earth (figure 12.1)—they show that our current discussions of synthetic biology are narrowly focused on the very near future. As they describe in their chapter, they were inspired in their work by the artist and philosopher Manuel DeLanda, who asks us to imagine an incredibly large temporal perspective, where instead of rocks we see flows. From this standpoint, the Himalayas would become a ripple in the surface of Earth, and species change would seem perpetual.[1] These ideas encourage a humbling shift in our perception of synthetic biology, and of all human activity for that matter, which becomes irrelevant or even insignificant from the perspective of geological time.

It was not just the geological timescale that interested Oron and Hideo. They used time as a critical lens on synthetic biology across a range of different temporal scales. As they explain, an inspiration for their focus on time was the organism that Hideo studies: cyanobacteria. Like all bacteria, cyanobacteria have incredibly rapid metabolic processes, but they are unusual because they are one of the simplest organisms to have circadian rhythms, or "biological clocks." This means that they have day–night cycles. Another distinctive feature of cyanobacteria is that they photosynthesize, giving them a green color (which is why they are sometimes known as blue-green algae). Most organisms have rapid metabolic processes and many have circadian rhythms (including all plants and animals), but cyanobacteria are particularly interesting because they also operate on much longer timescales. They can trap minerals from water and in the process build so-called living rocks, known as thrombolites and stromatolites (figure 12.2), as discussed in the previous chapter. These living rocks grow less than 1 millimeter per year and can develop over thousands of years. Even more impressively, cyanobacteria are incredibly significant on the geological timescale, as they are thought to be the organisms that originally introduced oxygen into Earth's atmosphere through their photosynthetic activities, making life possible as we know it. So cyanobacteria are relevant at timescales stretching from milliseconds to millennia.

One of the first things Oron and Hideo became interested in was manipulating cyanobacteria at the metabolic timescale, the scale at which most of synthetic biology operates. Metabolically, cyanobacteria are potentially very useful for synthetic biology because they can be used to produce high-value products such as alternatives to plastics, and they can play a role in wastewater treatment and in the production of biofuels, food, and fertilizers.[2] These types of application are typical of current synthetic biology; they are focused on the near future, with hopes of "quick wins" and significant new industrial developments within the next decade.[3]

Figure 12.1
The geological timescale puts
all human activity, including
synthetic biology, in perspective.

Figure 12.2
Cyanobacteria photosynthesize and deposit minerals, forming thrombolites or "living rocks," shown here at an early stage of development.

Oron and Hideo explored the idea of engineering cyanobacteria to digest the silicon in discarded computer chips, perhaps secreting the gold that is embedded in them. They noticed that one of the products of the breakdown would be fool's gold (iron pyrite), which looks like gold but lacks the properties we value in the precious metal, such as malleability and resistance to corrosion. This prompted them to ask: What would happen if we used synthetic biology to produce something useless instead of something useful? This is a challenging question, and it is the kind of question that is unlikely to be asked by scientists and engineers working in the field. It uses the metabolic timescale to critique synthetic biology in its current form. It immediately focuses our attention on the often unchallenged claim that synthetic biology will deliver products that will be beneficial for society in the near future. It also raises broader questions about the value of artistic and scientific work that does not have an obvious utility, an important theme for this pair.

Oron and Hideo's focus on time also encourages us to consider history and to ask questions about the historical progenitors of synthetic organisms. Some commentators have argued that synthetic biology will produce life forms that will be new in an important sense, as they will lack a causal connection to the historical evolutionary past.[4] This, they maintain, will make them significantly different from genetically modified organisms, which can be traced back to their "parent" species. But others argue that there is actually a continuum. They point out that much synthetic biology uses well-known organisms, such as *Escherichia coli*, which clearly have an evolutionary history. They maintain that even those branches of synthetic biology that aim to produce novel biologies that are completely independent of or "orthogonal" to existing biologies (like "more rational" versions of the DNA backbone) still draw on building blocks that can be found in nature, such as nucleic acids. Some work even attempts to allow synthetic biology to go back in history— *Jurassic Park*–like—by resynthesizing putative ancient proteins based on our current understanding of the geological record.[5] All this raises important questions about what it means for synthetic organisms to have a history. What counts as a break with the past? When are continuities more important? And does a focus on history make us think of synthetic biology differently?

As these questions show, Oron and Hideo's bold historical sweep draws attention to important issues. Rather than presenting us with a defeatist conclusion about humanity's inevitable demise, it forces us to think about the future and the kind of responsibility that we hold toward it. Their temporal perspective is an incredibly humbling one. It reminds us that all human activities will leave a legacy, and perhaps in synthetic biology's case a living legacy. Their work not only encourages humility, it also jolts us from the anthropocentrism of the "now" to take responsibility for the future. If cyanobacteria can already deposit rocks that last for thousands of years, what could synthetic cyanobacteria produce? If iGEM teams considered the existence of future landfill sites for discarded synthetic biological products, would this change the projects they chose to do? If professional synthetic biologists designed with future fossils in mind, would they design in the same way?

Actively pursuing the issue of time also brings into focus important questions about what makes something alive. Oron and Hideo's work suggests that we should think of living things as necessarily temporal, as fluctuating and changing. It shows the connections between time and life. This was an important issue for them, because they were concerned with what they perceived to be the extension of an atemporal "engineering logic" into the realm of living things.

Whether or not there is a single "engineering logic" in synthetic biology is something that can be debated (as we have seen in previous chapters of this book), but there are some features that all engineering-influenced approaches share. Perhaps most importantly, engineers aim to improve the world by developing useful products. To do this, they need to be able to manipulate and control the substances they work with. But manipulation and control are not the only ways of interacting with the biological world, and if engineers are the dominant voice in synthetic biology, we may expect to see the growth of a certain type of synthetic biology aligned with this particular agenda. Oron and Hideo's aim was to question this "engineering logic" and provide a contrast using the ambiguity and irony of artistic expression. Their work shows that there are many different approaches and perspectives that we can take, temporal and otherwise, to think about scientific and technological developments. If scientists, engineers, artists, and designers joined forces, this could broaden current understandings of the potentials and consequences of synthetic biology in ways that currently cannot even be imagined by any of these groups alone.

Art and Critique

Whereas some of the Synthetic Aesthetics projects are about design, this project explicitly aimed to produce art. Unlike other Synthetic Aesthetics residencies, where the artist and scientist had completely different skill sets, in this pairing Hideo, a biologist, is also a practicing artist, and Oron, a bio artist, has had his own laboratory for many years. In this sense they are both hybrid people. Hideo currently has five resident artists in his small lab at Waseda University of Tokyo, and the artists and scientists share the same benches. Oron has made a unique institutional niche for himself at the University of Western Australia by developing SymbioticA, "an artistic laboratory dedicated to the research, learning, critique and hands-on engagement with the life sciences."[6] Both Oron and Hideo want to use art as a way of critiquing synthetic biology. But what does this mean? How can art be a form of critique of synthetic biology?

On a very straightforward level, engaging with a scientific field as an outsider to that field will, almost by necessity, bring a critical perspective, which is critical just by virtue of being different. This is a form of critique that synthetic biology, as an interdisciplinary field, has to engage with on an almost daily basis. But art can bring a new dimension to this interdisciplinary mix because it provides a platform to explore important issues, such as the nature of life, which may be sidelined by the focus required by day-to-day scientific research. Hideo argued that that one of the strengths of art is that it can link different spheres and orders of experience; that it can combine in new and unexpected ways things that are usually separated. We see this in their collaborative project, where the bringing together of geological and biological time gives rise to a host of challenging questions for synthetic biology.

An important way in which art, particularly bio art, can be used for critique is by using the very same technologies and materials as are used by the scientists and engineers, but doing something different with them. The topic of analysis is also the medium of expression. Using the tools of synthetic biology to produce art means that this type of engagement is highly participatory because it requires a good understanding of the science. It also provides an opportunity for people with totally different backgrounds and assumptions from scientists and engineers to engage in a very similar hands-on experience. In this way, bio art makes use of the equipment, tools, and organisms that biologists use "but configures them in ways that try to provoke, transgress or re-design our understandings of life."[7] This is a critical step because it subverts the uses for which these tools and techniques were originally intended.

Art and Design

Oron and Hideo were adamant that their project should be an art project and not a design project. So I want to end this chapter by raising a challenging question provoked by this particular collaboration: What are the differences between art and design, and what does this difference mean for the way in which artists and designers engage with synthetic biology?

As previous chapters have explained, synthetic biology aims to make biology into a new material for design. It promises to redesign existing organisms, or design completely new ones. The type of design that is normally discussed in synthetic biology is engineering design, often in the context of the "engineering design cycle." The aim of engineering design is to develop a technical solution to solve a particular problem. Design in this sense aligns with the engineer's ameliorative impulse and fits well with Simon's (1969) definition of design as the "transformation of existing conditions into preferred ones."[8]

There is a multitude of definitions of design, of course. According to some, "Design is about service on behalf of the other."[9] In this sense a successful design is one that fulfills the desires of the client. This is not necessarily true for art, which is not usually driven by the satisfaction of external goals. Other definitions of design emphasize the importance of function: a product that is designed is designed to have a particular function.[10] From this perspective, we can legitimately ask questions about whether something is designed well or not, with respect to the function that it is designed for. Art, in contrast, can lack clear utility,[11] but it can nevertheless perform an extremely important social role. For example, cyanobacteria that produces fool's gold is useless from an engineering perspective, but not from an artistic perspective, because it makes us think about the purpose of synthetic biology. But this discussion raises a broader question: If art does not aim to develop useful applications, and synthetic biology does, then how should an artist engage in synthetic biology?

Hideo and Oron are dealing with this question by developing objects and ideas that explore the ambiguities and ambivalences of synthetic biology. Oron maintained that while designers (and engineers) are trained to find solutions to problems that bring closure, for artists the intention is rather to produce contestable objects that stimulate discussion. Rather than being prescriptive, these objects require those who encounter them to make their own decisions about technology and its place in society. Oron and Hideo's aim is to produce something that would make the observer stop and reassess his or her prior assumptions. They want their work to be part of a critical discussion, which does not have a defined endpoint, but is characterized by continuous exploration. For these reasons, they see their collaborative work as art rather than design.

But the line between art and design is blurred by movements such as "critical design," a school of design that developed in the 1990s[12] and can be seen as continuous with the radical design movement that arose out of Italy in the 1960s.[13] As discussed in chapter 3, critical designers use design as a tool to make us think about the social, political, and economic complexity of both technology and design. Possible technological futures are explored through speculative objects that provoke debate by making abstract ideas tangible and open to discussion (examples in this book include *The Scatalog* in chapter 6 and the packaging that creates its own contents in chapter 9). As Paola Antonelli has noted, the aim of critical design is not to solve a problem, but to challenge our assumptions about the role of technology in everyday life. These designers argue that this process of putting forward alternative technological futures is a form of critique.

Critical designers Anthony Dunne and Fiona Raby maintain that their work is not art because it is closer to the everyday and appears in the form of a potentially feasible product.[14] They argue that its everydayness means it has greater power to disturb than art, which, because it is often shocking and extreme, can be dismissed more easily as irrelevant. Another reason why it can be argued that their work belongs under the heading of "design" is that it is directed toward a user—although that user is often thought of as a "citizen" rather than a "consumer."[15] In developing imaginary products to stimulate debate, critical designers have to put themselves in the position of the users of these products and think about how these speculative products would fit into their lives (figure 12.3).

The critical design approach clearly has important similarities to that taken by Oron and Hideo. Critical designers, like artists, aspire to open up

science and technology to broader perspectives and to explore the ambiguities that emerge in a nonprescriptive manner. Both groups produce work that is sometimes uncomfortable and disturbing. Both critical design and art aim to expand our ways of imagining synthetic biology, which is also an aim of this book.

An important difference, however, is that Oron and Hideo are not concerned with designing speculative products for imaginary users. Because critical designers assume the existence of a user of synthetic biological products, they assume that the field will exist in the future. From Oron and Hideo's temporal perspective, this is not necessarily the case. Critical designers forgo the ultimate criticism of synthetic biology that is possible for artists: that it will not be a successful technology.

This difference in critical orientation between artists and designers was exhibited by Oron, whose relationship to the Synthetic Aesthetics project was

Figure 12.3
Between Reality and the Impossible (2010) is an example of Dunne & Raby's critical design work, which speculates about the future uses of technologies. Here, *Future Foragers* use synthetic biology to extract nutritional value from non-human foods.

always ambivalent. He was concerned that the project, because of its focus on design, would result in a beautification of synthetic biology that would promote the field. But despite these concerns, he thought it was important that artists have a voice in the discussion of synthetic biology, so he decided that it was better for him to be involved than to comment from the periphery as an outsider.

We can turn Oron's concerns on this book itself. It is an object that has been designed, and we, the authors, have attempted to make it into a beautiful object. In doing this, has the book lost its capacity to be critical of synthetic biology?

By thinking about synthetic biology from a temporal perspective, this project enables an exploration of the interconnected themes of utility, design, art, and critique. Oron and Hideo showed how time can be used as a critical lens on synthetic biology at many different temporal scales, from milliseconds to millennia. This critical lens encourages humility and responsibility toward the future, as well as recognition of the importance of history. It raises questions about utility, about design, and about the value of artistic and scientific work that serves no external purpose. Most importantly, Oron and Hideo's artistic engagement with synthetic biology forces us to think in ways that may be unfamiliar and uncomfortable and challenges our narrow preoccupation with the short term.

Structure

13.
Synthetic Sound from Synthetic Biology

Chris Chafe and Mariana Leguia

Sonification is a process by which nonaudible information is conveyed in audible form. Sonification of biological data is not a new idea. Others before us have taken elements derived from biology, such as the genetic code carried in molecules of DNA, and turned them into audible pieces. Approaches to re-codify the genetic code through sonification are as varied as the motivations driving the people that embark upon such endeavors. However, the one thing that remains constant throughout these exercises is that sonification enables a way to experience the genetic code that is dramatically different from the traditional "visual" one. Since its discovery, the genetic code has been written as a linear succession of combinations of Gs, As, Ts, and Cs used to represent the four nucleotides of DNA (guanine, adenine, thymine, and cytosine, respectively). Thus, molecules of DNA are "seen" rather than "heard." By re-codifying the genetic code to be heard rather than seen, we open the door to novel ways of experiencing and analyzing it, and ultimately to rethinking it.

Our collaboration intended to take sonification of biological data one step beyond a simple re-codification/sounding of the genetic code. To achieve that, we focused on the concept of structure. Structure permeates every level of biology, to the point where without it, there is no life. The function of proteins, for example, is completely dependent on their structure. If they are not folded in the correct three-dimensional shape, they will not be able to function as they normally do. In music, structure is also essential, and it significantly influences how we experience sound. The sound of one "klunk," for example, can be different from that of a "klank," and that quality is something to which we as humans are sensitive, even when we are not consciously aware of it. Structures built up of sounds and organized in time are the essence of music. Thus, the re-presentation of structure as a temporal flux is intended to touch our built-in sensitivities: whether probing sound qualities as when we knock on different materials to hear their structure or organizing sounds into musical arrangements.

Here, we explore how microscopic structures can be used to generate macroscopic sounds and, additionally, how these sounds can be used in "music" and represented in three dimensions in order to highlight qualities of structure and space. We have experimented with two separate approaches to sonification. Both draw from computer music techniques that Chris uses on a regular basis in the course of his work, and in both cases, structures were derived from various arrangements of DNA sequences that Mariana created through her work on the automation of DNA fabrication techniques for synthetic biology. In the first, we use elements contained within circular molecules of DNA to compose an ordered sequence of sounds that is then played in way that mimics the structure of the original circular molecule. In the second, we borrow from simulations of concert hall reverberation to explore alternate ways of creating sounds derived directly from structure itself.

Orbiting DNA Code with a Gear Shift

Central to synthetic biology is the manipulation of the genetic code, which often requires the stitching together of disparate pieces of DNA in order to create genetic circuits with particular functions. These genetic circuits can be assembled to do a variety of things, including turning genes on or off or modulating expression of other circuits. Many of these genetic constructions take the form of circular molecules called plasmids. Plasmids are naturally occurring structures, frequently present in a variety of bacterial species. Their function is to carry genetic material that is not normally part of the organism's main chromosome. Plasmids contain sections of DNA that encode distinct functions. Thus, specific segments can be grouped together into "families" according to the function(s) they encode. Families of sequences can encode genes that are expressed, regulatory regions that are used to turn genes on or off at particular times, and so on. The simplest plasmids contain sequences that enable a living cell to make more copies of the plasmid (known as ORIs—for origins of replication), sequences that enable finding that particular plasmid among others (known as selection markers), sequences that are of particular interest (YFG—for "your favorite gene"), and sequences that dictate the conditions under which YFG will be turned on or off given a specified set of parameters. Every year, researchers around the world build thousands of these plasmids during the course of routine research activities.

Mariana's work involves developing methods for the automated construction of DNA plasmids. As part of that work, she built a series of iterations (500+) of a simplified genetic circuit using variations of sequences belonging to different families. Essentially, every iteration had the same overall structure (a single circuit containing two genes, each marked with a specific recognition tag), but the sequences (or "flavors") representing those structures varied (two different genes and 12 different tags were used). In every case, however, she ended up with a circular plasmid that shared similar structural elements with all others in the set. Given the large number of plasmids constructed, and given that the genetic code is usually examined visually, comparing and contrasting iterations for quality control was a labor-intensive process. As part of this collaboration, we set out to explore the possibility that "hearing" DNA constructions might be a better alternative for parsing than visualization.

Our first approach has been a somewhat conventional interpretation of sonification, at least in its initial stages. From a compositional point of view, it has produced a 6-minute demo that is a musical study in advance of future gallery installation work. Specifically, we took a circular plasmid and examined it for structural elements, such as ORIs, selection markers, YFGs, and regulatory regions, that would be amenable to re-codification. We then sonified these elements in a variety of ways (figure 13.1). Finally, we played them in a circular arrangement, mimicking the structure of the plasmid from which they were derived, by using a ring of sound created with eight loudspeakers. The azimuth of a sound source was made to correspond to the position of each structural element around the plasmid structure. That is, sound came from a location in the circle of speakers that corresponded to a similar location in the circle of the DNA plasmid. The sound was then made to spin around the circle at different rates.

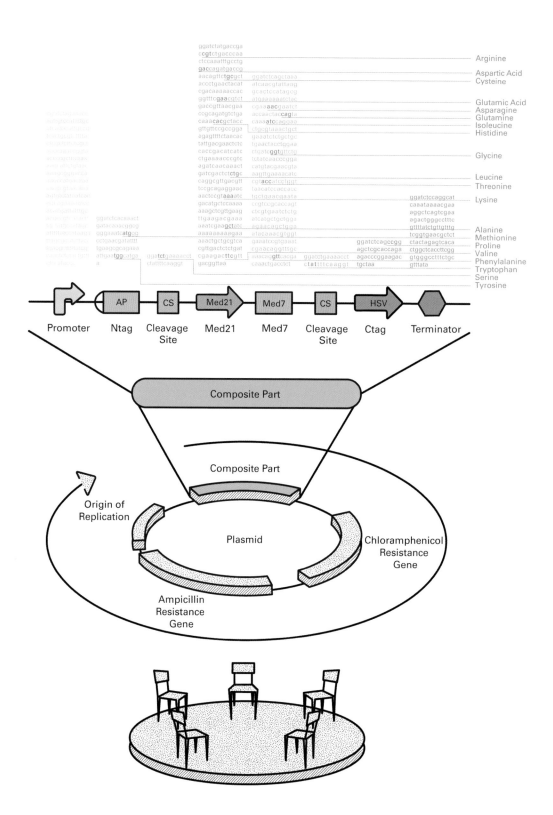

The composition simulates a gear shift going up and down its gears. At low speeds, simple microstructures, such as the linear succession of nucleotides that compose the plasmid are apparent. The listener can hear the plasmid nucleotide by nucleotide. As the composition speeds up, recordings of Mariana's voice are used to introduce additional audible labels that represent elements in the various families of synthetic biology parts. At the fastest speeds, only the ring's macrostructure, corresponding to elements encoding functions, is apparent. In fact, at this level individual nucleotides are played so fast that they almost blend into noise. When the composition slows down again, various levels of microstructure reemerge. The demo descends through various zoom levels corresponding to different "gears." At the beginning, an audience listening in the middle of the circular arrangement of speakers experiences the widest level of zoom, or the highest gear, as a fast "swish" of sound that orbits around. This swish is a noisy sound composed of individual sounding nucleotides, each of which triggers a brief musical tone, but because the sound moves around quickly, individual sounds per se are not easily detectable and instead blur together into a continuum. At this initial speed, each rotation around the entire plasmid, which is about 10,000 nucleotides in circumference, takes about 1 second to complete. A change in the orbiting speed marks movement through various zoom levels, and as the gears shift, individual nucleotides become perceptible as a (still rapid) series of notes, but which now move very, very slowly around the circle. The orbiting speed around the circle can be expressed as "angular velocity," and the ratio of angular velocities from the fastest "swish" to the slowest motion (in which individual DNA "notes" are distinct) is of the order 100:1. The demo traverses these extremes, lingering at different angular velocities.

At intermediate levels, we added the effect of "voice labeling" so that one can hear the names of structural units. For example, slowing to a certain speed, one hears the names of functional parts spoken, such as "promoter," "terminator," and so on. With these spoken labels, the listener perceives the location of the part within the circle. At an even slower ("zoomed in") speed, the names of individual amino acids, such as "alanine" and "isoleucine," are spoken. These are the building blocks from which biological parts are made. The "gear changes" visit several levels of abstraction by slowing or speeding the rate at which the plasmid is scanned musically, and these shifts are repeated in the course of the demo music. When the musical rhythm slows to its most extreme "ritardando" (deceleration), the demo reaches a point at which individual percussive notes are sounding "tick, tock, tack, . . . tuck, tack, tock, . . . tock, tock, tack," and we are hearing sounds representing the four nucleotides in the plasmid DNA sequence. That is, C, G, A, and T are represented by "tick," "tock," "tack," and "tuck." At this level of representation,

Figure 13.1
A plasmid (a circular piece of DNA) can be represented in many ways. At the top, we see the most common—a string of letters that stand for each nucleotide in the plasmid. At the top right is a list of the amino acids those nucleotides code for. In the middle, we see the diagrammatic representations used in synthetic biology, which represent the plasmid as a series of "parts." At the bottom is our sonic representation. The plasmid is "played" to listeners, who can control what they hear by manipulating the speed of the playback. They can hear each individual nucleotide played as a different note, or hear the names of amino acids, or hear the names of functional parts.

sounds come grouped as three-note sequences in order to highlight the fact that three nucleotides encode a given amino acid. The demo ends with a final zoom all the way out to the widest level, at which the plasmid itself is again represented as a fast swish around the circle. It is at this level that we might imagine a future piece of music contrasting different plasmids as musical motifs, objects, or themes that can be replicated, manipulated, and which can themselves in turn become building blocks for even higher levels of abstraction that are outside the scope of the current demo.

The fact that Chris, as composer of the music, might make musical constructions using operations on the data that resemble what Mariana, as synthetic biologist, might make with her engineered plasmids is to us the most exciting aspect of the work. Synthetic biologists work at many levels of abstraction, particularly when cutting and pasting discrete stretches of genetic material that encode particular functions. The goal of this sonification demo is to create an appreciation of the various levels that are superimposed and that can be derived from seemingly simple underlying detail. By analogy, this would be similar to giving a computer programmer an audible taste of the dense, intricate machine code used to execute programs written in a high-level programming language.

Our goal, stated literally, concerns the "re-presentation of structure as a temporal flux." From Chris's side of the project, the work involved a new kind of application of extra-musical data to music. Where previous work in sonifying/musifying data always involved time series data, Mariana's structures in this project were static. That meant finding a time element outside of the data that could have musical importance. Ultimately, the hoped-for musical engagement is created by the gear shift itself, which allows navigation of the structural features. We envision taking this beyond the current demo, which is something of a "drive through," to a future interactive setting such as a gallery installation in which there can be an infinity of paths taken through the structures. Each visitor's exploration of one of Mariana's molecular structures will be a new realization of "the piece" in sonic terms.

DNA Rooms

One of the first things a guitar maker will do is to knock on and listen to the wood from which he or she intends to create a new instrument. The quality of that reverberation will be integral to the sound the guitar is later able to produce. Knocking or tapping on individual molecules or strands of DNA is not part of the routine in most biology research labs. Nevertheless, recent advances in technology are pushing the envelope of what is now possible. For example, the atomic force microscope physically taps on single molecules that are splayed out on surfaces. Similarly, high-resolution optical "tweezers" can pull on individual strands of DNA, tightening and relaxing them, and doing so repeatedly until they snap. The fact that it is now possible to observe and probe biophysical properties of molecules directly at the atomic level made us wonder if we could also find ways to "listen" to them, such as by plucking a stretched strand of DNA as if it were a guitar string. Soon, we discovered a different approach to probe a variety of sound structures encoded in single molecules of DNA. Our approach produces sounds that relate to the subtle way in which we perceive the acoustics of spaces we inhabit by listening to patterns of echoes. For example, the echoes in a large church or concert hall are dramatically different from those in a small room. Similarly, the echoes from a particular room will vary depending on whether it is empty or filled with objects. In computer music, reverberation echoes can be created to mimic room structures. In turn, these can be imposed on sounds to give the impression that they are being played in synthetic, imaginary rooms.

A contrast of perspectives is noteworthy here. Chris, as a musician, talks in terms of materials and/or spatial structures, that is, things that have "mass" and/or "volume," whereas Mariana talks of structure in the sense of a linear representation of nucleotides, or parts, or genes, where mass is less relevant and instead what is interesting is the patterns contained within the linearity of the molecule.

We have exploited a similarity between the chromatograms produced by Sanger-based DNA sequencing and the time-domain patterns of echoes in rooms. Chromatograms are visual representations of sequencing data corresponding to a segment of genetic code, whereas time-domain graphs of echo patterns are used to see the myriad of sound reflections that a listener hears when making any sound in a room. The latter phenomenon is somewhat like a billiards table on which a ball, being struck, makes many bounces off walls before coming to rest (figure 13.2). Of relevance to this project is the fact that both signals can be drawn as two-dimensional graphs of similar shape (figure 13.3). Initially, both signals show large amplitudes, but eventually the intensity of the signals decrease. In the case of chromatograms, fluorescence intensities representing each nucleotide along a DNA sequence decrease as one moves away from the sequencing initiation point (figure 13.4), whereas in the case of time-domain impulse responses, it is the sound pressure level of the room echoes that decreases as time goes on. In computer music, recorded echo patterns (impulse responses) are routinely used as building blocks to make complex room simulations. Given the observed pattern similarities between chromatograms and impulse responses, we attempted to construct room echoes from DNA by extracting the information needed from Sanger-based sequencing chromatograms (figure 13.5). In looking together at these

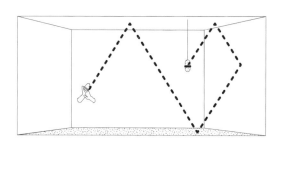

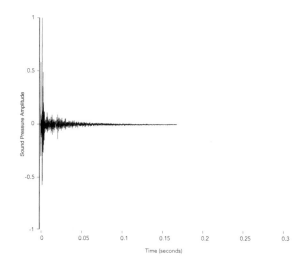

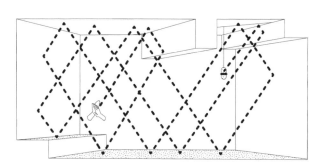

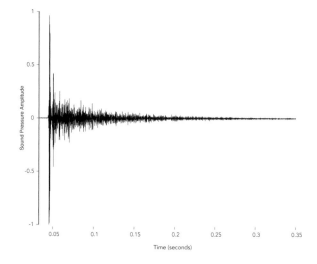

Figure 13.2
Different rooms produce different sound reverberations. A clap will "bounce" around a room and sound a particular way to listeners because of the shape and arrangement of the walls and any objects present in the room.

Figure 13.3
The graphs shown are real diagrams of sound reverberations, illustrating how two rooms—the first larger than the second—shape the acoustic experience in different ways.

graphs (figures 13.4 and 13.5), each of us interprets something different that is dictated by the norms in use within our respective fields of work. However, the ambiguity and the similarity between the two types of data suggested that, with the tools available at our individual laboratories, one a synthetic biology wetlab and the other a computational music lab, we could create synthetic reverberation from synthetic biology.

Synthetic reverberation is often made by sampling echoes of real rooms. All that it takes is an impulse like a clap or a balloon pop and a recorder to pick up the train of echoes that makes up the impulse response. Usually, impulse responses are quick and sound like a brief "knock" or tap on a desk. Other times, they can be drawn out and take a long time to die out, as in the case of the reverberation produced in a concert hall or cathedral. Regardless of whether reverberations are fast or slow, we are sensitive to their different qualities even if we do not understand the acoustics and mathematics needed to describe them. Thus, a young child will be able to sense intuitively a real difference between a clap produced in a bathroom and the same clap produced in a concert hall. Moreover, a fascination with sound and sound environments is something that children develop at the youngest age. They cannot help vocalizing when being wheeled in their strollers through a reverberant tunnel or passageway.

Because Sanger-based data decay just like an impulse response, nucleotide sequences can be reinterpreted as sound pressure levels that sound convincingly similar to the pattern of echoes that might be recorded from a room. In other words, they can be turned into a brief sound that can be used as the sonic "fingerprint" of a room's structure. In this manner, different DNA sequences can be used to make different impulse responses as if recording different echo patterns obtained by clapping in different rooms. Using software and convolution reverberation techniques,[1] chromatograms can be turned into "room claps" and then used to reverberate recordings of instruments and voices in unique-sounding synthesized rooms. That is, music and voices can be played as if reverberating inside a "DNA room." Specifically, we made six contrasting DNA rooms derived from chromatograms corresponding to distinct DNA sequences. The chromatograms were used to reverberate synthetic salsa music, created by Jacob Wittenberg, an undergraduate computer music student at Stanford. Jacob was researching automatically generated music of different styles and had succeeded in producing "never ending" salsa rhythms made of different drum sounds. An algorithm of his own design created interesting and varied performances and had been used earlier in a collaboration between Chris and Greg Niemeyer. *Tomato Quintet* was an installation space where visitors would hear music controlled by CO_2 levels modulated by ripening tomatoes (at the end of the installation, a salsa eating and salsa listening fest was held). His salsa music

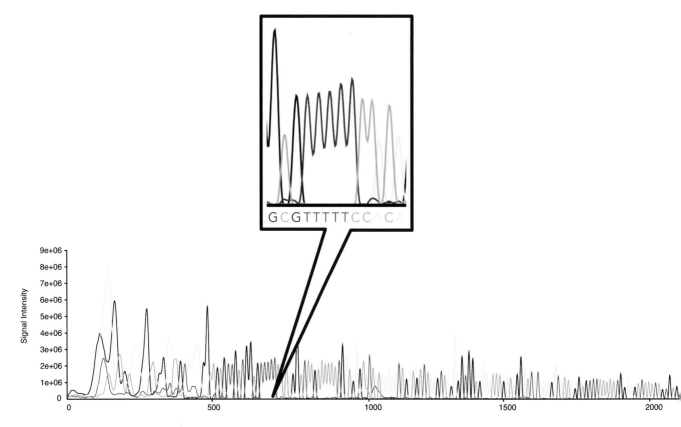

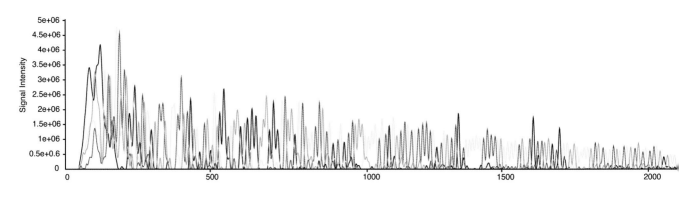

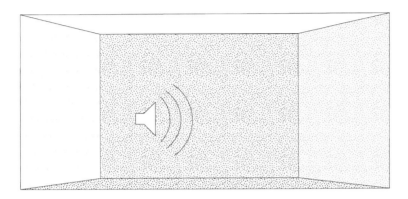

2500   3000

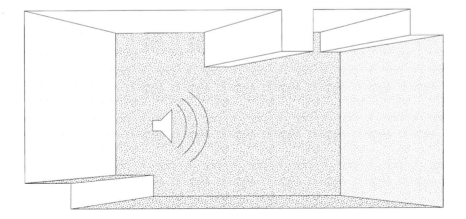

2500   3000

Figure 13.4
Two different DNA sequences.
When DNA is sequenced, the
result is a chromatogram. We
noticed that chromatograms look
similar to sound reverberation
traces and decided to make
virtual "DNA rooms," represented
in figure 13.5.

Figure 13.5
We played back music as if it were
being played in a room whose
acoustic properties are defined by
the chromatogram data of a piece
of DNA (figure 13.4). What would
that room look like?

was synthesized "dry" such that it had no room sound to start with. To give it a bit of "flavor," we processed a sound clip of Jacob's salsa music with six different DNA impulse responses and produced six independent sound files. With each recording, one hears the same musical source played in different acoustical environments. The results range from bathroom acoustics to club stage, and although some feel more natural than others, all generated plausible rooms. We believe that this is a somewhat lucky result given that there is obviously no inherent correspondence between DNA and acoustical "space." And it could easily have turned out otherwise, for example if the DNA pattern grew again in strength midway before decaying to nil. It would be an "un-room-like" behavior to have energy not continuously decay.

Whether turning DNA sequences into "room sounds" will have any practical application is doubtful at the moment but ultimately remains to be determined. Regardless, the point of this collaboration was not to produce something with inherent practical application, but rather to begin to think in creative ways that may serve as seeds for other bigger and better ideas borne out of unlikely collaborations. As unlikely partners as synthetic biology and computer music may be, it is undeniable that both fields are lively, creative, and, perhaps surprisingly so, share elements in common that can serve as foundations to speak and create in common.

14.
Abstraction
and Representation

Pablo Schyfter

Even the most cursory look at synthetic biologists' writings will soon run into the notion of "abstraction," and for good reason: The concept plays a key role in how the field thinks about and hopes to work with the living world. In general, we tend to think of "the abstract" as the domain of absentminded philosophers and Modernist art. "The abstract" tends to be portrayed as that which stands opposed to the real world, either by way of idealized concepts or bizarre, otherworldly artwork. And yet, in synthetic biology "abstraction" has all to do with putting real-world things together; it is about making the engineering of living nature happen. Importantly, it is about making possible a type of working by way of a particular view of the world. That is, "abstraction" is all about representation.

"Representation" is very important in thinking about the Synthetic Aesthetics project and the particular case of Chris Chafe and Mariana Leguia's work with sound. Unfortunately, representation is like abstraction, insofar as it is a difficult concept to pin down: It can refer to many different things. American pragmatist philosopher John Dewey writes, "To say in general that a work of art is or is not representative is meaningless. For the word has many meanings. An affirmation of representative quality may be false in one sense and true in another."[1] That is to say, "representation" has no single, definitive meaning. In the most straightforward of senses, a painting can depict a landscape or a sculpture can render the likeness of a famous individual. However, these simple examples of artistic representation do not exhaust representative practice in the arts. A painting that does not portray the likeness of a place or object may still act as a depiction of emotions, ideas, intentions, values, or desires. Dewey argues that architecture is representative insofar as it expresses the "memories, hopes, fears, purposes, and sacred values"[2] of those who build edifices. A building can represent the "enduring values of collective human life."[3] As such, to say that art represents is not to say that it "merely" reproduces things as accurately as possible, as "representation" is a rich constellation of practices.

Colloquially, we use "representation" to mean "standing in for" or "serving as a portrayal of" something or someone. We are represented by our elected officials in democratic government, and a painting can represent a landscape or an individual. Although the term serves countless other uses—particularly a variety of technical senses in the humanities and social sciences—this vernacular understanding of "representation" works surprisingly well here.

"Representation" is about replacing one thing for another. A stop sign represents a command and a legal regulation. An elected official stands in for a constituency and (ideally) acts as a substitute for a community of voices. From this starting point—representation as "standing in for"—we can begin to think about abstraction.

When synthetic biologists refer to "abstraction," they point to a way of thinking about biological things.[4] Sequences of DNA become "parts." When those parts are put together, they form "devices" or "circuits." When devices themselves are combined, a "system" is the result. Once this system is good-to-go, it can be inserted into a "chassis," a term used to refer to the host cell. Words and ideas taken from the world of mechanical and engineered things are used as stand-ins (representations) for biological stuff.

Terms like "part" and "device" stand in for bits of nucleic acid. Importantly, they stand in *imperfectly*. To say that such-and-such a bit of DNA is a "part" obscures much of what makes DNA a particular kind of thing. That is, some of the specificity of DNA is lost. This is not so much a criticism as it is a note about the way things are: representation is never perfect; something of the original is always lost.

The very same is true of abstraction. After dusting off and looking through a few dictionaries, I find that the meaning of "abstraction" includes overtones of withdrawal, seclusion, and independence from the qualities of the real thing. When we abstract something, we pull away from it. The "real," out-there thing is left behind as we step away toward the abstraction, and our view of the thing is transformed by that withdrawal.

To abstract is to represent. It is to make use of a stand-in idea. In synthetic biology, the goal of this representation is engineering practice. In art, representation can serve innumerable ends. Chris Chafe and Mariana Leguia use sound and music to interrogate abstraction in synthetic biology. As I see it, they use representation to question representation. Their exploration of synthetic biology is compelling because it makes use of artistic representation to critique a particular type of scientific representation: engineering-based abstraction. Chris and Mariana's contribution suggests some important questions. When synthetic biologists use abstraction, what is left behind? What is lost, and how does that loss affect our relationship to the living world?

Representation and Scientific Business-as-Usual

Science represents. More accurately, science makes use of representation; scientific business-as-usual uses a host of representational tools and techniques in constructing knowledge about the natural world.[5] Again, there is no single method of representation, nor is the most obvious the most important. It is tempting to discuss scientific representation solely in terms of visual reproductions. After all, scientists use a range of imaging techniques: some straightforward and others much more elaborate. I am epileptic, and what goes on inside my skull has been represented over the years by electroencephalograms, by CT scans, and by MRI scans. It is not all that difficult to understand these processes and tools as representational. However, there are other, less explicit ways in which science represents: data and metaphor.

Many of the sciences are intensely quantitative activities. That is, they use numerical information and mathematical theories to make claims about the world. Through experimentation, an object or an event can be translated into a series of numbers—it becomes data. Data stands in for the actual thing or phenomenon: it is a representation. Following our understanding of representation, data stands in and substitutes. Of course, data does not

```
gatctctatgctactccatcgagccgtcaattgtctgattcgttaccaattatgacaacttgacggctacatcattcacttttcttcacaaccggca
cggaactcgctcgggctggccccggtgcatttttaaataccgcgagaaatagagttgatcgtcaaaaccaacattgcgaccgacggtggc
gataggcatccgggtggtgctcaaaagcagcttcgcctggctgatacgttggtcctcgcgccagcttaagacgctaatccctaactgctggc
ggaaaagatgtgacagacgcgacggcgacaagcaaacatgctgtgcgacgctggcgatatcaaaattgctgtctgccaggtgatcgctgat
gtactgacaagcctcgcgtacccgattatccatcggtggatggagcgacctgttaatcgcttccatgcgccgcagtaacaattgctcaagcag
atttatcgccagcagctccgaatagcgcccttccccttgcccggcgttaatgatttgcccaaacaggtcgctgaaatgcggctggtgcgcttca
tccgggcgaaagaaccccgtattggcaaatattgacggccagttaagccattcatgccagtaggcgcgcggacgaaagtaaacccactggt
gataccattcgcgagcctccgagcgtacgaccgtagtgatgaatctctcctggcgggaacagcaaaatatcaccctggcggtcggcaaacaaattct
cgtccctgattttttcaccaccccctgaccgcgaatggtgagattgagaatataacctttcattcccagcggtcggtcgataaaaaaatcgagat
aaccgttggcctcaatcggcgttaaacccgccaccagatgggcattaaacgagtatcccggcagcaggggatcattttgcgcttcagccatac
ttttcatactcccgccattcagagaagaaacaattgtccatattgcatcagacatttgccgtcactgcgtctttttactggctgcttctctcgctaaccaa
accggtaaccccgcttattaaaagcattctgtaacaaagcgggaccaaagccatgacaaaaacgcgtaacaaaagtgtctataatcacggca
gaaaagtccacattgattatttgcacggcgcgtcacactttgctatgccatagcattttttatcataagattagcggatcttacctgacgctttttatc
gcaactctctactgtttctccataccggatctcacaaactgatacaaacggcgggaaatcatgggcctgaacgcgattttttgaagcgcagaa
aattgaatggcatgaaggatctgaaaacctctattttcaaggtggatctatgaccgaccgtctgacccaactccaaatttgcctggaccagatg
accgaacagttctgcgctaccctgaactacatcgacaaaaaccacggtttcgaacgtctgaccgttaacgaaccgcagatgtctgacaaacac
gctaccgttgttccgccggaagagttttctaacactattgacgaactctccaccgacatcatcctgaaaaacccgtcagatcaacaaactgatcga
ctctctgccaggcgttgacgtttccgcagaggaacaactccgtaaaatcgacatgctccaaaaaaagctcgttgaagttgaagacgaaaaaat
cgaagctatcaaaaaaaaagaaaaactgctgcgtcacgttgactctctgatcgaagacttcgttgacggttaaggatctcagctaaaatcaac
gtattaaggcactccatagcgatgaaaaatctaccgaaaacgaatctaccaactaccagtacaaaatccaggaactgcgtaaactgctgaaa
ataagcacaagtttttatccggccttattcacattcttgcccgcctgatgaatgctcatccggaatttcgtatggcaatgaaagacggtgagctg
gtgatatgggatagtgttcacccttgttacaccgtttccatgagcaaactgaaacgtttcatcgctctggagtgaataccacgacgatttccg
gcagtttctacacatatattcgcaagatgtggcgtgttacggtgaaaacctggcctatttccctaaagggttttattgagaatatgtttttcgtctc
agccaatccctgggtgagtttcaccagttttgatttaaacgtggccaatatggacaacttcttcgcccccgtttttcaccatgggcaaatattatac
gcaaggcgacaaggtgctgatgccgctggcgattcaggttcatcatgccgtttgtgatggcttccatgtcggcagaatgcttaatgaattaca
acagtactgcgatgagtggcagggcggggcgtaatttgatatctgagctcgcttggactcctgttgatagatccagtaatgacctcagaactc
catctggatttgttcagaacgctcggttgccgccgggcgtttttttattggtgagaatccaagcctcggatctctgaagactctcgagctgcagg
actcacagctcctcggatctcttagacgtcaggtggcacttttcggggaaatgtgcgcggaaccccatttgtttattttttctaaaatacattcaaat
atgtatccgctcatgagacataaccctgataaatgcttcaataatattgaaaaaggaagagtagtgagtattcaacatttccgtgtcgcccttatt
cccttttttgcggcattttgccttcctgtttttgctcacccagaaacgctggtgaaagtaaaagatgcagagatcagttgggtgcacgagtgg
gttacatcgaactggatctcaacagcggtaagatccttgagagttttcgccccgaagaacgttttccaatgatgagcacttttaaagttctgcta
tgtggcgcggtattatcccgtattgacgccgggcaagagcaactcggtcgccgcatacactattctcagaatgacttggttgagtactcacca
gtcacagaaaagcatcttacggatggcatgacagtaagagaattatgcagtgctgccataaccatgagtgataacactgcggccaacttactt
ctgacaacgatcggaggaccgaaggagctaaccgcttttttgcacaacatgggggatcatgtaactcgccttgatcgttgggaaccggagctg
gaatgaagccataccaaacgacgagcgtgacaccacgatgcctgtagcaatggcaacaacgttgcgcaaactattaactggcgaactactta
ctctagcttcccggcaacaattaatagactggatggaggcggataaagttgcaggaccacttctgcgctcggcccttccggctggctggttta
ttgctgataaatctggagccggtgagcgtgggtctcgcggtatcattgcagcactggggccagatggtaagccctcccgtatcgtagttatct
acacgacggggagtcaggcaactatggatgaacgaaatagacagatcgctgagataggtgcctcactgattaagcattggtaactgtcaga
ccaagtttactcatatatactttagattgatttaaaacttcattttaatttaaaaggatctaggtgaagatcctttttgataatctcatgaccaaaatcccttaacg
tgagttttcgttccactgagcgtcagaccccgtagaaaagatcaaaggatcttcttgagatccttttttttctgcgcgtaatctgctgcttgcaaacaaa
aaaaccaccgctaccagcggtggtttgtttgccggatcaagagctaccaactctttttccgaaggtaactggcttcagcagagcgcagataccaa
atactgttcttctagtgtagccgtagttaggccaccacttcaagaactctgtagcaccgcctacatacctcgctctgctaatcctgttaccagtggc
tgctgccagtggcgataagtcgtgtcttaccgggttggactcaagacgatagttaccggataaggcgcagcggtcgggctgaacggggggtt
cgtgcacacagcccagcttggagcgaacgacctacaccgaactgagatacctacagcgtgagctatgagaaagcgccacgcttcccgaaggg
agaaaggcggacaggtatccggtaagcggcagggtcggaacaggagagcgcacgagggagcttccagggggaaacgcctggtatctttat
agtcctgtcgggtttcgccacctctgacttgagcgtcgatttttgtgatgctcgtcaggggggcggagcctatggaaaaacgccagcaacgcgg
cctttttacggttcctggccttttgctggccttttgctcacatgttctttcctgcgttatcccctgattctgtggataaccgtattaccgcctttgagtga
gctgataccgctcgccgcagccgaacgaccgagcgcagcgagtcagtgagcgaggaagcggaagagcgcccaatacgcaaaccgcctctc
cccgcgcgttggccgattcattaatgcagctggcacgacaggtttcccgactggaaagcgggcagtgagcgcaacgcaattaatgtgagttagc
tcactcattaggcaccccaggctttacactttatgcttccggctcgtatgttgtgtggaattgtgagcggataacaatttcacacaggaaacagcta
tgaccatgattacgccaagcgcgcaattaaccctcactaaagggaacaaaagctggagctgcaagcttaatgcggtagtttatcacagttaaat
tgctaacgcagtcaggcaccgtgtatgaaatctaacaatgcgctcatcgtcatcctcggcaccgtcaccctggatgctgtaggcataggcttgg
ttatgccggtactgccgggcctcttgcgggatatcgtccattccgacagcatcgccagtcactatggcgtgctgctagcgctatatgcgttgatg
caatttctatgcgcacccgttctcggagcactgtccgaccgctttggccgccgcccagtcctgctcgcttcgctacttggagccactatcgactacg
cgatcatggcgaccacacccgtcctgtggatcctctacgccggacgcatcgtggccggcatcaccggcgccacaggtgcggttgctggcgccta
tatcgccgacatcaccgatggggaagatcgggctcgccacttcgggctcatgagcgcttgtttcggcgtgggtatggtggcaggccccgtggc
cgggggactgttgggcgccatctccttgcatgcaccattccttgcggcggcggtgctcaacggcctcaacctactactgggctgcttcctaatgc
aggagtcgcataagggagagcgtcgaccgatgcccttgagagccttcaacccagtcagctccttccggtgggcgcggggcatgactatcgtcg
ccgcacttatgactgtcttctttatcatgcaactcgtaggacaggtgccggcagcgctctgggtcattttcggcgaggaccgctttcgctggagc
gcgacgatgatcggcctgtcgcttgcggtattcggaatcttgcacgccctcgctcaagccttcgtcactggtcccgccaccaaacgtttcggcga
gaagcaggccattatcgccggcatggcggccgacgcgctgggctacgtcttgctggcgttcgcgacgcgaggctggatggccttccccattat
gattcttctcgcttccggcggcatcgggatgcccgcgttgcaggccatgctgtccaggcaggtagatgacgaccatcagggacagcttcaagg
```

need to be numerical (figure 14.1). The three-letter sequence "UAG" can be used to represent a particular biological object called a stop codon. Instead of giving the full chemical description for this bit of RNA, it can be summarized by the use of this alphabetic representation: three letters stand in for a chemical structure.

In a similar manner, metaphor can stand in for reality. When a biologist compares DNA to a computer program, she is using the concept of software as a substitute. DNA is not really a computer program, but it may be useful to think about it as such when carrying out research. Metaphorical representation is what abstraction accomplishes. To say that a DNA sequence is a "part" or that a cell is a "chassis" is to make an analogy. Instead of DNA, we now have nuts and bolts; instead of a cell, we now have a mechanical frame.

Figure 14.1
One way to represent DNA. Here, nucleic acids are indicated by one of four letters. This is a visual, rather than sonic, way to represent a DNA molecule.

Numerical representations can be used to formulate mathematical theories of the world. Alphabetic representations can serve as shortcuts. Metaphors and abstractions can help simplify a messy, complex world and highlight things of interest. Each type of representation, including abstraction, delivers some form of benefit to those who use it. If it did not, its use would be senseless. However, those benefits are not free of charge. Mathematical theories, analytic shortcuts, and simplifications come with a price. Something is always lost in representation.

Sonification

Chris and Mariana make use of a particular form of representation called sonification. As the name suggests, sonification uses sound. To come to terms with what sonification accomplishes, consider some simple, everyday instances of it. For example, think about the sirens of an ambulance or the error noises produced by a computer. In both cases, sound is being used to convey a message: respectively, "clear the road" and "you've attempted an invalid action." Sonification is a "planned" form of audial representation, but we also routinely take impromptu sound cues from the world around us. Thunder betokens a gathering storm, and a car engine that "doesn't sound right" suggests the need for a visit to the mechanic.

Sonification can also be used to serve artistic ends. Series of numbers can be translated into series of notes, which when played compose an acoustic representation of those numbers. For instance, a Stanford University class in electronic acoustics makes use of a data set of monthly dry white wine sales in Australia between 1980 and 1995. The number of thousands of liters per month can be assigned to different notes, the time between each month can be set at a particular interval of seconds, and the record of wine sales can be "played." Using this basic concept, musicians such as Chris—director of Stanford's Center for Computer Research in Music and Acoustics—translate nonacoustic phenomena into sound. By way of illustrating this work in more detail, consider one of Chris's recent projects. Working with neurologists at Stanford's School of Medicine, he translated electroencephalogram data (figure 14.2) from an epileptic patient into an acoustic composition.[6] The data, which represents the electrical activity of an epileptic brain, becomes a complex arrangement of sounds. One can hear the epileptic brain. Notably, this composition is a representation of a representation. That is, it is an acoustic rendering of electroencephalogram data, which in turn represents cerebral electrical impulses.

For the Synthetic Aesthetics project, Mariana—a synthetic biologist at the time working at the University of California, Berkeley—and Chris mobilized sonification in order to interrogate how abstraction is used in synthetic

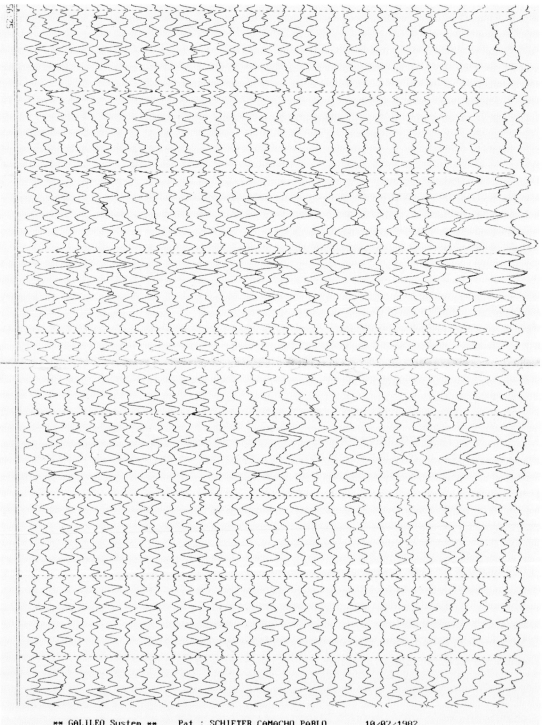

** GALILEO System **   Pat : SCHIFTER CAMACHO PABLO   10/07/1982
Record : 13/06/02 15:56:55 File relative time : 00:02:30
[S: 7 uV/mm T: 0.3s F: 30 Hz . M]

biology. The pair began by trying to match sonification to the tools and techniques of synthetic biology. Focusing on a plasmid—a circular piece of DNA—Chris and Mariana developed a number of different methods to represent genetic structures using sound. First, the pair considered using four different notes as stand-ins for the four nucleotides of DNA, and then "playing" the plasmid's genetic sequence using sound. They soon realized that this sequence is extensive, and much of it is of little interest to synthetic biologists; playing the whole genome might be unnecessary. Thus, the second composition only sonified the important sections—the genes of interest—while leaving the rest of the plasmid "silenced." This choice mirrored synthetic biologists' focus on particular "parts" and their corresponding overlooking of the remaining sequences. Chris and Mariana then asked themselves an interesting question: How exactly should these "important" sections be sonified? Simply playing the sequence of nucleotides might serve little use, and each "part" can be called by a number of different names. For instance, one might call something by its part type, by its colloquial shorthand name, by its technical name, or by the name of the chemicals coded for by the DNA sequence. This variety of different ways to identify a bit of DNA led the pair to its first encounter with the issue of abstraction. One can choose to "hear" each part in a number of different ways, and each of these ways both conveys information and withholds information. Knowing that a particular part is a "promoter" tells me what it does, but it doesn't tell me that it happens to be a particular promoter, say *lacP*. Then again, knowing that a part is *lacP* might mean nothing to me if I don't already know that *lacP* is a promoter. Different forms of abstracting from the DNA sequence serve different ends for different users and require different sets of preparatory knowledge.

To convey this multiplicity of uses and abstractions, Chris overlaid a variety of different sounds for each component of the plasmid. By controlling the angular speed with which one loops around the plasmid, one might choose to hear: the full DNA sequence; only the "important" sections of the DNA sequence; the important sections read out by their technical name; the important sections read out by their function. A listener working with these audial representations can decide what to hear about the plasmid and how to hear it. The listener controls the form and experience of representation, and those choices follow from whatever interests or concerns the listener. This simple fact tells us a lot about representation, abstraction, and synthetic biology.

Representation, Abstraction, and Design

In my chapter on design and nature (chapter 5), I argue that design is all about choices. Something similar can be said about representation. In setting something as a substitute or stand-in, decisions must be made about what will

Figure 14.2
Electroencephalogram data from an epileptic person. The lines represent electrical activity inside the brain.

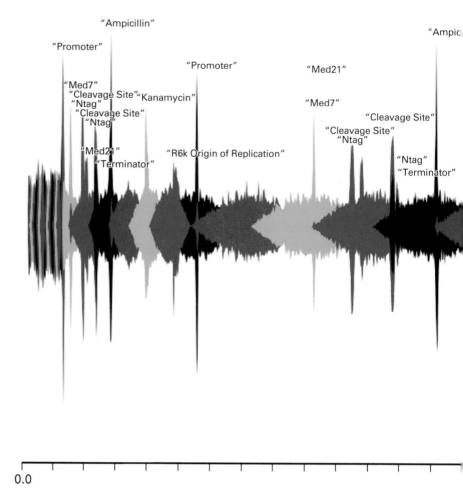

work best, and these decisions can be complicated and involved. As with design, representation does not just happen of its own volition. Representation is carried out, and it is carried out in specific ways determined by the type of people doing the representing (figures 14.3 and 14.4). It would be foolish to suggest that just because art and science both make use of representation, art and science represent for the same reasons and in the same ways. After all, art and science do not aim at the same ends. So, while both fields make use of stand-ins and substitutes and express things through proxies, thinking about what representation *actually does* in each field requires that we think about the specifics of those fields. In the case of synthetic biology, thinking about representation means thinking about abstraction.

Sonification and abstraction are both forms of representation. They differ in how that representation is carried out. Where sonification uses sound, abstraction in synthetic biology is all about introducing a simplifying concept

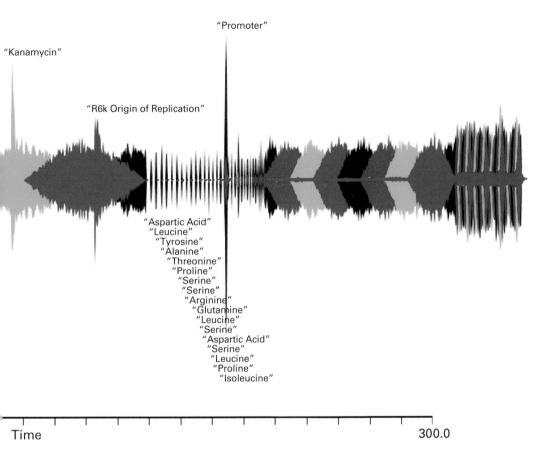

"Kanamycin"

"Promoter"

"R6k Origin of Replication"

"Aspartic Acid"
"Leucine"
"Tyrosine"
"Alanine"
"Threonine"
"Proline"
"Serine"
"Serine"
"Arginine"
"Glutamine"
"Leucine"
"Serine"
"Aspartic Acid"
"Serine"
"Leucine"
"Proline"
"Isoleucine"

Time   300.0

or analogy. Saying that a bit of DNA is a "part" or that a cell is a "chassis" makes the object easier to understand. There are a number of pragmatic reasons for doing so. Abstraction can make thinking about complex systems easier, and it may serve synthetic biology in its quest to practice engineering on living nature. Nonetheless, abstraction is not without its complications. Chris and Mariana, in bringing sound-based representations and engineering-based abstraction, challenge us to scrutinize what synthetic biology is doing when it abstracts the things of living nature with mechanical or electronic metaphors. Their work demands we ask: What does abstraction— a form of representation—accomplish in synthetic biology? How does it influence our view of and engagement with the living world? To explore these questions, we must return to the matter of engineering.

Synthetic biology is working to become an engineering discipline. A core goal of those who make up the field is ensuring that living stuff can be used

Figure 14.3
DNA made sonic. This diagram shows various levels of sonification produced by Chris and Mariana's work. The colored pulses correspond to nucleic acids represented by sound. The words correspond to genes represented by Mariana's voice.

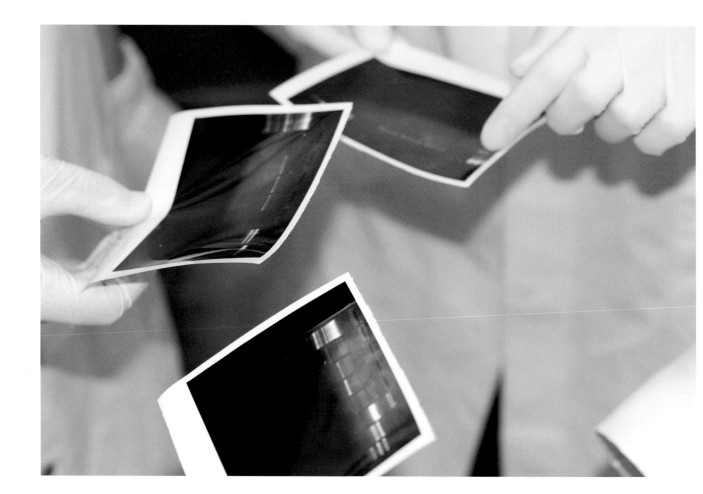

Figure 14.4
Images produced by gel electrophoresis. The white lines and blocks are used to represent portions of DNA. Here, lines and blocks "stand in" for molecules.

to design and build technological things. The tools and techniques of the field—including its use of abstraction—are geared to this end: How can we design and how can we build with biology? Representation in synthetic biology—in the form of abstraction—serves the making of technological things. So ultimately, representation is important as something that gives shape to how we interact with the living world.

Look back to chapter 1. Figure 1.6 in that chapter is an example of how synthetic biologists represent the stuff with which they work. The "parts," "devices," "systems," and "circuits" I discussed earlier are represented by arrows, circles, boxes, and computer programming terms. At first glance, the image appears to be one to do with electronics. There is nothing about it that would immediately indicate that it stands in for a sequence of pieces of DNA, just as overhearing two people speak about "parts" and "devices" would not lead someone to conclude that the conversation is about biological things. And yet, for those initiated in the language of synthetic biological abstraction, diagrams of this kind can be quite useful: They make thinking about the DNA construct easier; and they may serve as broad guides for building DNA constructs. A synthetic biologist looking at such an image might be able to understand what the pieces of DNA are intended to do and deduce how the construct should be arranged. For experts in the field, a diagram of that kind—a representation of a bit of DNA—is a design tool and a blueprint for construction.

In many ways, this type of representation is very successful. It does what it's supposed to do: help the practical aim of making technological things. By abstracting away from the messiness of the real-world thing, a diagram may help make engineering with biology possible. If we are satisfied with this, our discussion can perhaps end there. But Chris and Mariana press us to ask some further questions: What gets lost or left behind in this type of representation? From what do we pull away?

Loss and Withdrawal

Representation is about standing in or substituting. Obviously, no stand-in is identical to that which is being represented. Elected officials are not indistinguishable from their constituents and often exercise judgment contrary to what those constituents might wish. The most accurate artistic representation of something can never be identical to what it portrays. A photograph of a landscape captures that place from only one perspective at only one point in time, and it fails to capture the smells and sounds of the location or the feel of a breeze experienced by the photographer when taking the shot. Representation always involves some form of transformation: some things are left behind, others changed, and others still introduced. Artists bring idiosyncratic perspectives on what they represent—views unique to the individual. In fact, we expect them to do so. In science, natural things and events are represented in terms that serve theoretical explanation, and this may involve considerable modifications. Images in journals often use false-color reproduction to highlight things of interest. Mathematical data is a radical transformation of the world into a form that serves quantitative theories. In both artistic and scientific representation, things are routinely lost, set to the side, or added.

Abstraction, as a form of representation, is subject to this dynamic. But as I noted earlier, abstraction also involves withdrawal. Abstracting something involves pulling away from that something. The abstract concept of "justice" may serve philosophical speculation well, but it does not carry the nuances of writing laws or the contingencies of courtroom argumentation. This is not to say that abstraction is a wrong done to real-world things; it is simply to acknowledge that the abstraction stands apart—sometimes far apart—from what is abstracted. Now, take the case of synthetic biology. From what does abstraction in synthetic biology withdraw? When a DNA sequence is represented by way of a diagrammatic sketch of pseudo-electronic circuitry, what is left behind?

In demonstrating the many ways that the "same" object can be represented, Chris and Mariana press us to ask these types of questions. Each representation will enable certain practices and curtail others. As part of that, each representation will carry with it a body of assumptions and aims. Speaking of "parts" and "devices" in synthetic biology—abstracting away from the chemical bits of DNA—cannot be divorced from the belief that what has worked for us in the realm of metal, rock, and plastic will serve us equally well in making with the stuff of living nature. What is left behind is the specificity of living things *as living things*.

An organism is a particular kind of thing for many reasons. One of these is its quality as something that is alive. That quality is not one shared by the materials routinely used by engineers, and it is precisely the quality pushed aside when synthetic biologists carry out their abstraction. In chapter 8, I make the case that metaphors can obscure the biology of biology: those things that make living nature unique, different, and valuable. That chapter and this one ask similar kinds of questions. Metaphors can hide the same type of specificity lost in abstraction. Ultimately, I have no position against the use of metaphors and abstractions in synthetic biology, as long as those tools and practices are viewed as postulates, rather than foregone conclusions. Synthetic biologists should not be satisfied to think about living things as stuff little different from hardware store merchandise, particularly when doing so turns a blind eye to the "living" in "living things."

My point here is not advocacy for some type of eco-mysticism. In fact, I am motivated by something fairly straightforward. Synthetic biologists already recognize and praise the uniqueness of living nature; they just happen to do so *selectively*. Again, even the most cursory look at synthetic biologists' writings will reveal that many find the distinctiveness of biology compelling. If this is the case, shouldn't attempts to build with biology begin from a place of acknowledging that distinctiveness?

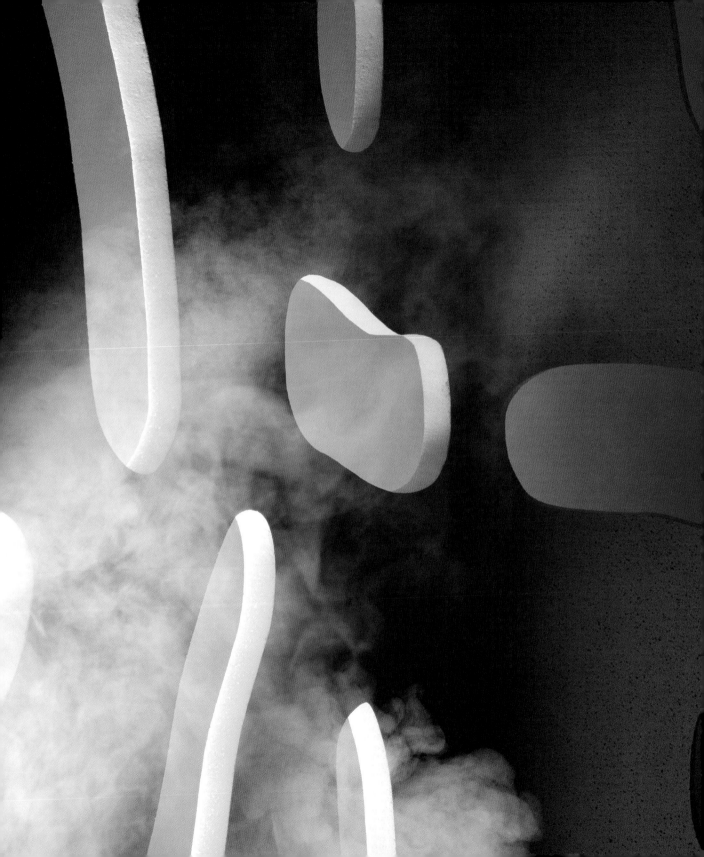

Evolution

15.
Living Machines

Sheref Mansy
and Sascha Pohflepp

Throughout much of recent history, technological advancement has led to the replacement of animal technologies with inanimate, human-made machines. Now, in the second decade of the twenty-first century, one of the dominant narratives of progress is that of a "century of biology." A movement is gaining momentum that increasingly looks at life as a "technology" that, if properly engineered, can be used to construct living or at least life-like machines. Here we will trace a history of the inanimate machine and explore whether what we now view as a traditional machine is in fact just an interlude in engineering history. The hypothesis presented is that current thinking and future technologies may come to allow for the merging of machines and engineered life into a new, third form. These living machines would have radically different and surprising properties but may also integrate better with the environment than any previous artifact of the industrial world.

Animal, Mechanism, Machine

At first, the concept of a living machine may seem absurd, because machines and life appear distinct. Machines are rationally designed by us to carry out desired tasks. Life, however, came into existence neither by the intentionality of humans nor for the benefit of humans. Yet, it could be claimed that some of the first machines, that is to say tools that significantly enhanced people's abilities to alter the environment, were indeed alive or had living components. Living organisms such as food plants and farm animals have had a long history of use, in one way or another, as objects that consume energy to produce an output, not unlike the much more recent inanimate machines. As Clay McShane and Joel Tarr point out in their essay about the decline of the horse in the American city, "understanding that the body (whether human or animal) was a site for energy conversion [was] the key insight of modern physiology."[1] Some of the first examples of use of animals as machines thus revolved around animals that were fed (input of energy) in order to harness their bodily movement (output) for transport. Similarly, plants have long been exploited for their ability to consume an energy source that we cannot directly exploit (sunlight) and turn it into a foodstuff that we can consume. In fact, because humans cannot physiologically produce all of the molecules required for survival, we are dependent upon the ability of other organisms to produce essential nutrients. Essentially, such early uses of organisms were already, in a sense, exploiting metabolism as a technology.

Breeding and selection were clearly used throughout history to improve the performance of animals, but technology continued in other directions as well. Over time, the utility of organisms was increasingly expanded by the incorporation of mechanical parts. Perhaps the invention of the plough at around 6000 BCE marks the beginning of a new era, which would eventually transform agriculture, transport, and warfare. The role of the organism, whether human or another animal, in these hybrid technologies simply was to power the system (figure 15.1). The generated functional whole, consisting of animal and inanimate parts, was perceived as a natural combination of body and mechanism that "converted the linear, ambulatory, slightly rhythmic gait of the animal to the rotary motion required by most machinery, usually by gears."[2]

If, for the sake of argument, we allow ourselves a wholly materialist perspective on the human use of living organisms, it appears that at that time, a contemporary notion of the machine may have been one that included contraptions that used a living body and its metabolism as an engine to drive

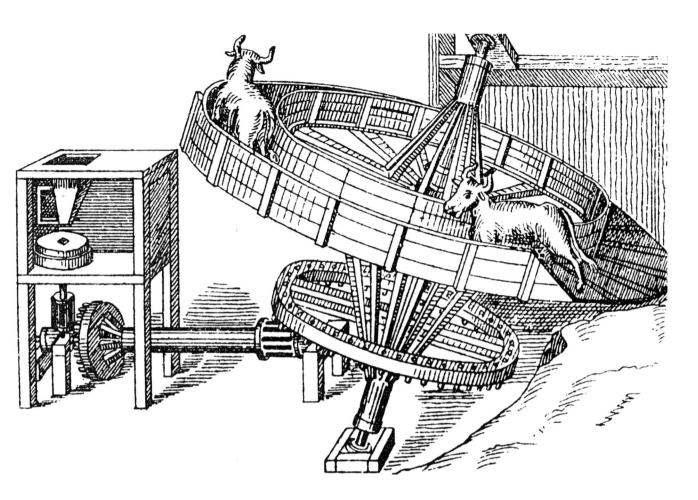

a mechanical process. With the Industrial Revolution, in many countries such machines were superseded by wood-powered or fossil fuel–powered, nonliving machines that, nevertheless, modeled living machines (figure 15.2). In other words, machines were built that substituted the animal engine with a nonliving mimic. In time, many tasks such as the threshing of crops were accomplished by mobile machines that were not based on animal power. Economic forces ensured that nonliving machines would almost completely replace animal power in the late nineteenth century and in the process shape our contemporary notion of a machine. As horses and oxen disappeared from the streets and farms of industrialized nations, we began to change our definition of a machine. Still, almost as if to demonstrate how relatively new our notion of a strictly nonliving machine is, as late as 1914 a book was published with a chapter entitled "The Horse as a Machine."[3]

Further evidence of the complex relationship between humans, mechanisms, and animals is evident in our contemporary vocabulary. For example, the power generated by nonliving machines is oftentimes quantified in

Figure 15.1
Horse or ox mills convert the bodily power of animals into a rotary motion that can be used to drive a mechanism such as a millstone.

Figure 15.2
Modeled on the bodies of horses
and oxen, agricultural steam
engines relatively swiftly replaced
their living counterparts.

horsepower, a unit originally intended to make numerically comparable horses and the engines that were in the process of replacing them. Indeed, the horse appears to sit at a symbolic point in the history of the machine. Not only did the demise of horse-powered machinery mark a shift toward the use of nonliving machinery, but also these new machines caused a change in exploited fuel sources. For example, animal-powered machinery typically consumes living fuel sources, such as crops, whereas nonliving machines run on nonliving fuels, such as wood, coal, gas, or oil.

This shift from living to nonliving machines may give us a hint of what is to come. After a period of "dead" machines, we may again see the emergence of living or life-like technologies. Similarly to how "horse-using societies unleashed their scientists and engineers to study equine machines and the horse became a form of technology,"[4] we may currently be witnessing a similar moment in which scientists and engineers study and build technologies with sequences of DNA. In this proclaimed "century of biology," despite or possibly because of the tremendous advancements achieved in building nonliving technologies, many scientists and engineers appear to be returning to the idea of life as a machine.

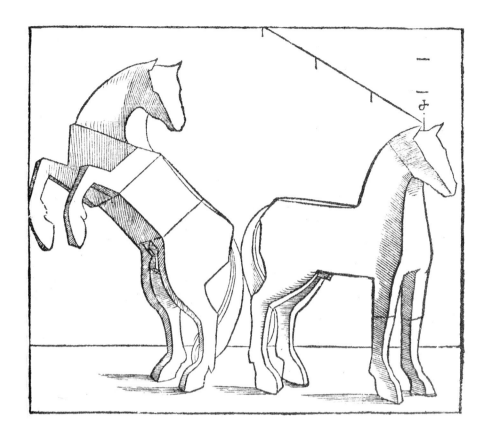

Bridging the Concepts of Life and Machine

The design and construction of life in ways that could never be achieved without intentional human intervention has come to be called "synthetic biology," although it has had other names before. We tend to think of such endeavors as something new. The reality is that the dream of engineering life is old. In fact, as Daniel Margocsy explains in "Designing the Horse,"[5] there is fascinating evidence for how horse breeders in the sixteenth century used geometry and regarded themselves as "engineers—not necessarily engineers of the human soul, but engineers of the animal body," all within the religious framework of a static, divine creation (figure 15.3). What is indeed new today are the tools at our disposal, including the theoretical tools for reimagining our role within the systems of the world and the practical tools for manipulating the material basis of life.

One of the conceptual tools that began to shift our view of life back toward something that is capable of being engineered may have been the science of cybernetics, a branch of interdisciplinary science concerned with the structure of regulatory systems. As described by the American mathematician Norbert Wiener in his 1948 book entitled *Cybernetics, or Control and*

Figure 15.3
In medieval Europe, printmakers such as Albrecht Dürer und Erhard Schön sought to construct the ideal horse purely from geometric shapes.

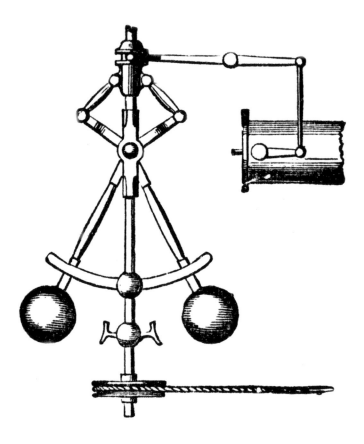

Communication in the Animal and Machine, cybernetics is a theory in part based on the study of the mechanisms of machines controlled through corrective feedback, such as James Watt's steam engine. This device, the very machine for which the unit horsepower was first applied, was equipped with a so-called "governor," a rotating assembly of weights mounted on arms that determined how fast the engine shaft spins and then uses proportional control to regulate its rotational speed (figure 15.4). Inspired by this regulatory relationship, the highly diverse proponents of cybernetics hypothesized that such regulatory designs are pervasive in nature and may be applicable to many fields stretching from computer science to engineering, anthropology, psychology, and, notably, biology.

Cybernetics changed the way in which nature is perceived by explicitly incorporating the regulatory systems of machinery into the descriptions of natural biological pathways. This cultural feedback loop, however, has not been without criticism. Scholars such as molecular biologist and science historian Lily Kay suspected cybernetics to be an attempt at a "reductionistic mathematization"[6] of life, as a deeper understanding of universal regulatory mechanisms also implies that we would eventually be able to engineer and

Figure 15.4
The production of steam in James Watt's engine forms a feedback loop that is regulated by a centrifugal governor.

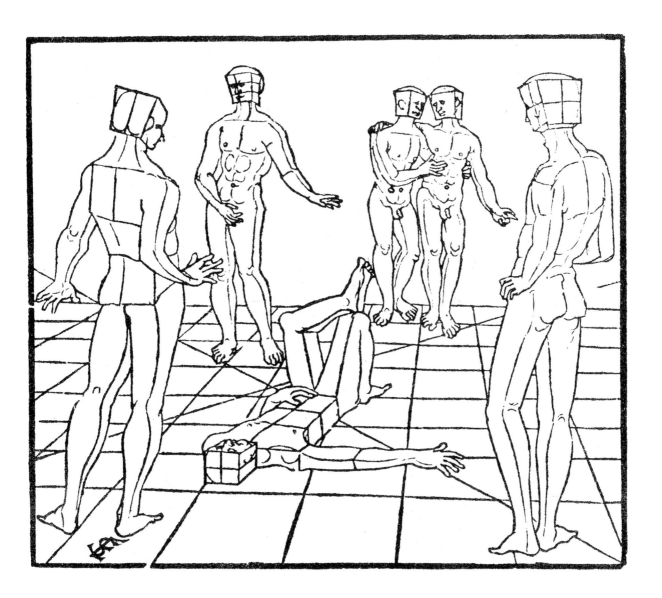

govern living systems ourselves (figure 15.5). Yet, within the evolution of a potential living machine, this moment in history appears to be crucial, as the theory of cybernetics along with its implications may have bridged the concepts of life and machine more profoundly than the functional combination of ox and plow or the breeding of a geometrically "perfect" horse could ever have done. It set the stage for a modern-day attempt at what we now call synthetic biology by suggesting that the living machines of agriculture, the nonliving machines of industrialization, our current information-age machines, and the coming new forms of living machines are in many ways one and the same.

Figure 15.5
In 1542, printmaker Erhard Schön explored the quantifiability of the human figure through geometry.

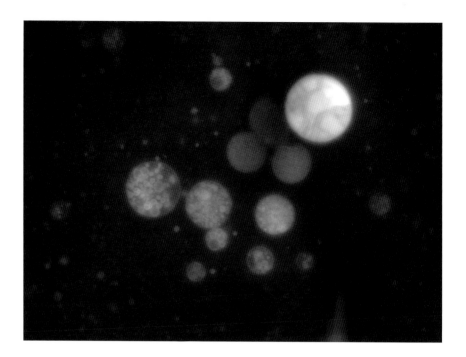

Contemporary Synthetic Biology

Currently, the engineering of life comes in many forms, which we divide here into two broad categories: "top-down" and "bottom-up." The top-down synthetic biology approach uses existing organisms, such as bacteria (e.g., *Escherichia coli*) or baker's yeast (*Saccharomyces cerevisiae*), often referred to as the "workhorses" of synthetic biology. It modifies these organisms by inserting or removing genetic information that encodes specific functions. Conversely, the bottom-up interpretation of synthetic biology attempts to build living cells or life-like systems "from scratch," essentially converting chemistry into biology (figure 15.6). The bottom-up perspective began as a quest to trace the origins of life but has grown into efforts to generate new technologies.

Synthetic biology has already revisited many of the features of the living machines of the past. Just as animal-based technologies were fueled by crops and exploited metabolism to drive movement, early successes in synthetic biology have returned to crops as fuel and to metabolism as a manufacturing process. The most well-known examples are probably the engineering of bacteria and yeast that grow by consuming sugars and produce products of commercial value, such as pharmaceuticals and diesel fuels. These living machines are built by incorporating DNA sequences that code for a body and a range of natural and artificial parts. In some respects, modern-day living machines are not different from centuries-old manufacturing practices, such as those used to make bread, beer, and yogurt, yet the difference is

Figure 15.6
Vesicles made from simple lipids spontaneously form structures morphologically similar to living cells.

the extent to which human intervention has shaped the metabolisms of the microscopic organisms that are being used. In their vast potential for future applications, these genetically engineered, invisible cells promise to collapse all previous notions of the machine into one, as outlined by cybernetics in the 1950s, potentially changing many aspects of human society in the process.

Designing with Darwin

If the engineering of life becomes more commonplace, we may come to view life more similarly to inanimate machines. But, because of the very fact that they are alive, living machines have profoundly different properties. Through evolution, the primary driving force of life is survival. Attempts at engineering living or life-like entities to carry out desired tasks in effect are just adding functional appendages to systems inherently focused on the paramount goal of "not dying." Within this framework, life will always explore all available options to survive and reproduce regardless of human impositions. Living machines thus have an agenda that is more than the by-product of a human purpose. Perhaps bottom-up synthetic biology methods that do not exploit existing cell architectures can avoid the problems associated with the differing goals of engineered organisms and the humans that engineered them. That is, of course, only true if life can exist in the absence of its own agenda.

Living systems replicate and evolve whereas inanimate machines do not. For example, James Watt's steam engine was designed and constructed by humans. The machine "runs" and over the course of time fails. The only way a machine qualitatively changes from how it was constructed without intentional intervention is by breaking due to the wear and tear of use. However, a truly living machine with the aforementioned agenda moves quite differently through time. A living system only has to be constructed once, because life proliferates through replication. A living machine too can break—that is, die—but in the meantime it also grows, reproduces, and evolves.

Replication thus constitutes a radical departure from classical notions of elegance in design and engineering and the importance of standardization in manufacturing. A single living machine can rapidly give rise to billions of nearly identical copies that are nevertheless unique. It is rarely appreciated that this aspect of life, even engineered life, originates far from a human place, and that contingent effects will always greatly outnumber human intervention. In other words, a single person, an architect, engineer, scientist, and so forth, may be able to piece together an artificial cell, but even after one replication cycle that person must share authorship with the innumerable forces that drive Darwinian evolution.

Imagine how strange it would be to build a machine that evolves over time. Even if proper working conditions and selective pressures were put in place so that the desired traits of the organism would persist, the internal workings of the cell would inevitably be subject to change. This could lead to a bizarre sequence of events where engineers design and construct a precise system, time and contingency change the inner workings of that system, and finally scientists are called in to investigate how the current version works. Here again, synthetic biology resonates with cybernetic thinking. Sociologist of science Bruno Latour, for example, describes the "black-boxing" of technoscientific systems: "when a machine runs efficiently, when a matter of fact is settled, one need focus only on its inputs and outputs and not on its internal complexity. Thus, paradoxically, the more science and technology succeed, the more opaque and obscure they become."[7]

We have not yet reached a stage where synthetic biology has produced widely used "black box" machines. However, laboratory experiments confirm that living systems naturally evolve over time to a state better suited for their specific living conditions. In other words, the organism adapts, but we do not know how. For example, Richard Lenski's laboratory at Michigan State University has been monitoring the evolution of continuously grown cultures of *E. coli* since February 1988. Even in the absence of intentional human intervention, all of Lenski's bacteria significantly changed their genomes and their physiology over the past 50,000 generations.[8] Astonishingly, as early as 1880, Rev. Dr. William Dallinger recorded similar data resulting from a 7-year study in Sheffield, England. In this work, which Darwin hailed as "extremely curious and valuable,"[9] Dallinger demonstrated that the gradual raising of environmental temperature induced an adaptive change in the bodies of the simple, unicellular organisms that he used as his subject (figure 15.7). After 500,000 generations, organisms that grew at 15°C were able to tolerate 70°C. Another interesting aspect of this work is that the experimental setup contained a true cybernetic feedback system half a century before the publication of Wiener's cybernetics book. Dallinger's 100-year-old experiment thus may give us a hint of what future realities of designing for living machines may look like by exploiting directed evolution in addition to or in place of rational engineering.

Life and the surrounding environment are completely intertwined into a complicated system in which a change in one initiates a cycle of changes in the other. More specifically, life feeds off and shapes the environment. The environment influences what can and cannot survive. Essentially, this means that life necessarily changes the environment by its very existence into a different state that in turn influences whether or not the organism can persist. The engineers and designers of living machines thus need to take into consideration the evolution of the organism and the evolution of the environment as a single, coupled mechanism.

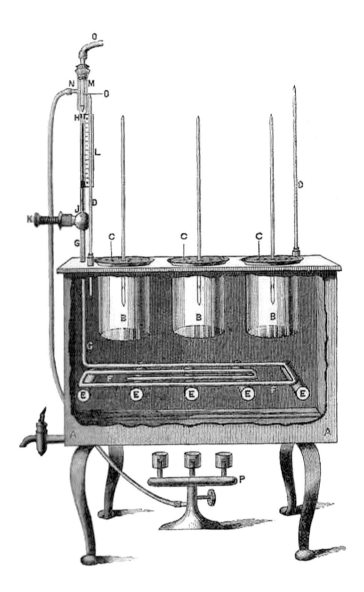

A Truer Feedback Loop

Could it be then, with the emergence of truly living machines and the appreciation of their radically different characteristics, that we are in fact partially returning to previous conceptions of life, machine, and the environment? Just as early innovations in machine technologies viewed the animal engine and its nonliving extension parts as a whole, we may be forced to consider a living machine and its environment as a discrete entity. Working with such a machine may feel radically different as it will possess the momentum of life. However, these living machines also would take us back to harnessing the power of a living body, not much different from the traditional harnessing of

Figure 15.7
The incubator used by Rev.
Dr. William Dallinger in 1887
gradually to raise the temperature
in his 7-year controlled evolution
experiment.

horse power. We would argue that this is a good thing, because a shift toward the use of living machines may allow for a more sustainable future.

The Industrial Revolution resulted in the vast use of fossil fuels as a temporal buffer that removed nonliving machines from the immediate energy cycles of the planet. Every combustion engine on Earth is running on borrowed sunlight from a distant past. One of the negative consequences of the Industrial Revolution has been the systematic modification of the environment toward a state that is less compatible with our existence. Industrial-age machines change the environment, but the environment can only hope to change the machines via the proxy of global politics or economics. Currently, this feedback loop is severely distorted. Perhaps living machines that evolve along with the environment will have the potential to strike a better, more dynamic balance.

16.
Evolution or Design?

Jane Calvert

A mule is an animal that combines the benefits of a donkey with those of a horse. Mules eat less than horses of their size, and they have stronger hooves. They are less obstinate and have more stamina than donkeys. They have been bred by humans who value their qualities and they have been used for centuries for farming and transportation. But the thing about mules, as every schoolchild knows, is that they can't reproduce. They are the result of a cross-species mating between a male donkey and a female horse (figure 16.1). Donkeys and horses have different numbers of chromosomes, so a mule has an odd number (63), which usually cannot pair up and create viable embryos. We breed mules for their useful

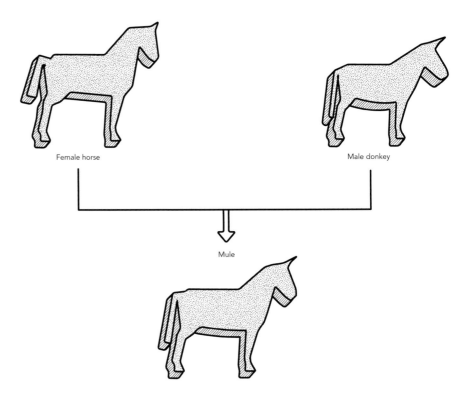

Female horse

Male donkey

Mule

Figure 16.1
Mules, which can't reproduce,
are the result of a cross-species
mating between a male donkey
and a female horse.

characteristics, but only for one generation. A mule might be described as a disposable biological system.

If synthetic biology succeeds in producing "living machines," will they be like mules—doing a job for one generation only? Or will they be more like horses—creatures that reproduce, evolve, and adapt to their environments? The answer to this question depends on what type of synthetic biology we are talking about. Some bioengineers might be happy with something that performs its job well, as a one-off. Protocell researchers, in contrast, are more interested in exploring and understanding the characteristics of living things through synthetic biology. They would want their synthetic creations to have the properties we normally associate with living things, like being able to reproduce, evolve, and adapt.

The point here is that there are several different approaches to synthetic biology, with different motivations and agendas. Much of the synthetic biology we discuss in this book is focused on developing applications by using biology as a material for engineering. But some synthetic biologists, including protocell researchers, are more interested in gaining new knowledge. They believe that a synthetic approach can be a powerful tool in furthering our fundamental understanding of the nature of living things. This is an aspect of synthetic biology that should not be overlooked.

Protocell researchers aim to create living cells "from scratch." They often start by making a simplified version of a cell membrane called a "vesicle," into which they may then put simple biological molecules.[1] Their long-term aim is to understand what life is, and they have strong connections to the "origins of life" field. The challenging question "What is life?" is often regarded as one for philosophical investigation, but protocell researchers attempt to address it through scientific experimentation. They hope that through mimicking the behavior of living things by building simple biological systems, they will gain a better understanding of what it is that makes something alive.

Unlike synthetic biologists working on standardized biological parts, protocell researchers do not usually see themselves as doing engineering. They do, however, share the engineer's aspiration to reduce complexity, saying that if they build an artificial cell, they want to understand how every bit of it operates, which means they must build it completely from scratch. Because of this desire for ultimate transparency and complete control, some protocell researchers argue that theirs is a more "authentic" form of synthetic biology than other approaches that work with pathways, genes, or genomes that already exist in nature.

As described in chapter 3, in May 2010 Craig Venter announced, with much fanfare, that he had created a "synthetic cell."[2] Most protocell scientists did not regard this as being a demonstration of the creation of life, however, precisely because it relied on preexisting biological material. Venter constructed an entirely synthetic version of a natural genome and implanted it into a recipient cell, where it took over the function of that cell and successfully replicated. But the synthetic genome only thrived because it was put inside an existing cell, which became a crucially important part of the new "synthetic cell." As one protocell scientist told me, what Venter had done was analogous to taking a house and putting in new furniture. Instead, protocell scientists want to build the house itself.

Protocell research initially appealed to Synthetic Aesthetics resident and biochemist Sheref Mansy because of the pioneering, exploratory nature of the work. He explained that a high degree of creativity is needed in this area, and that this requires exposure to different perspectives, such as design and art. The aim of protocell work is to expand our understanding of life, and his hope in joining the Synthetic Aesthetics project was that art and design could contribute to this goal. His ambition was to open up his work to different perspectives in order to improve its creativity and quality.

Sheref was paired with artist Sascha Pohflepp. Although Sascha was trained as a designer, he was clear from the start that he wanted to participate in the Synthetic Aesthetics project as an artist. He explained that this was because design is about finding solutions and solving problems within predefined constraints, whereas art is about posing questions rather than delivering answers (this resonates with the division made by Oron Catts, discussed in chapter 12). Sascha drew a similar distinction between engineering and science and argued that art has more in common with science than with engineering, because both science and art pursue understanding, albeit in different ways. This aligned well with Sheref's work, which he described as pure science.

There is a lively debate about the features that something must possess in order to be considered to be "alive," a debate I cannot do justice to here, but Sheref and Sascha's interests converged on the idea of evolvability as one of the defining characteristics of life. Their focus on evolution resonates with a prominent but simple definition of life from the U.S. National Aeronautics and Space Administration (NASA), which reads: "a self-sustaining chemical system capable of Darwinian evolution."[3]

Sheref and Sascha's joint work on evolution gave rise to many thought-provoking questions. They began by asking: If what makes living systems different from manufactured systems is the fact that they evolve, then what would this mean for the "living machines" that synthetic biology aims to produce? How would our notion of a machine (such as a clock or an engine) change in a future where many of our machines are biological? What would happen if nature became more machine-like and machines became more life-like? For example, what would happen to a steam engine (the icon of the Industrial Revolution) if it were allowed to evolve over generations? Would it rearrange itself and perhaps function better than it did before? Would this lead to a loss of control and a lack of understanding about how it actually worked? What commercial and industrial implications would this have? And how would undesirable characteristics that might evolve be dealt with?

Synthetic Biology and Evolution

Although these are important and interesting questions, the underlying issue I want to address here is the role of evolution in synthetic biology. If synthetic biology aims to design the living world, then how will it deal with the fact that one of the characteristics of living things is that they evolve?

Because there are different approaches to synthetic biology, there are different views on the role evolution should play. As we have seen, protocell scientists typically aim to build living systems that can evolve. Many other synthetic biologists aspire to make use of evolution in helping them reach their goals and draw heavily on what is called "directed evolution."[4] Still others think that evolution is best eliminated because it subverts our well-intentioned designs by changing them over time. Sheref and Sascha's project allows these different attitudes to evolution in synthetic biology to be explored and interrogated.

Directed evolution is a way of using evolutionary mechanisms in the laboratory to produce organisms with desired characteristics. It works in a very similar way to natural selection in the wild. The synthetic biologist starts with a mixed population of organisms and introduces a selection pressure that favors one type over another. In the wild, this selection pressure could be a predator, but in the lab it could be something like a high temperature. The organisms that are favored by the selection pressure replicate more than the others, and over several generations this type of organism comes to dominate.

Synthetic biologists often turn to directed evolution when they are confronted by a difficult problem that they cannot solve by purposeful design, because an advantage of directed evolution is that it can result in novel and unexpected beneficial changes to a system, which may not be the kinds of changes that a researcher would think to introduce. For this reason, some synthetic biologists say that evolution is the best designer we have, although they may admit that we do not yet understand completely how it can be best deployed. Some even describe themselves as "evolutionary engineers."[5]

Other synthetic biologists, particularly those with a background in engineering, argue that evolution is "tyranny" and that it needs to be eliminated (figure 16.2). They hope for a time where "sufficiently mature DNA sequencing and synthesis technologies will allow us to decouple the designs of life from the constraints of direct descent and replication with error."[6] The J. Craig Venter Institute's 2010 paper, mentioned earlier, went part of the way to fulfilling this objective by showing how the constraints of direct evolutionary descent can be overcome for a bacterial genome, which Venter's group showed could be made entirely synthetically.[7]

The problem with evolution from an engineering perspective is that things that evolve are things that change in unpredictable ways, and if the aim of synthetic biology is to design reliable and predictable biological systems that can be controlled, then we do not want them to evolve. But as evolution will act on anything that is replicating, it would seem that the only way to avoid this would be to make living systems that do not reproduce. To return to the metaphor used earlier, synthetic biologists would have to make mules rather than horses. A disadvantage of this approach is that it does not harness the reproductive potential of biology, which could potentially be very useful. As the synthetic biologist Tom Knight puts it: "Fundamentally, biology is a manufacturing technology. . . . The things it builds are more copies of itself."[8]

But it could be liberating to design a piece of biology without having to plan for its future evolution. For example, the human Y chromosome is the only chromosome that does not have a matching pair (it is paired with the much bigger X chromosome; figure 16.3). If something goes wrong

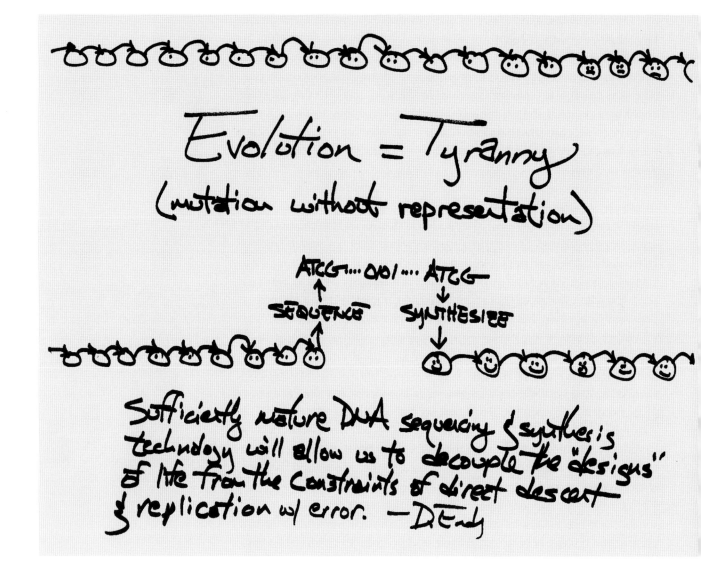

Figure 16.2
Drew Endy's "Evolution = tyranny" illustrates his hope that in the future, it will be possible to decouple synthetic biological designs from the constraints of evolution.

with the Y chromosome, then it cannot make use of the X chromosome. Instead, the Y chromosome has to use other parts of itself, so it has developed a very complex molecular structure of nested and inverted repeats.[9] This complexity means that errors are likely, and there is the danger of deletions, which can result in illness or infertility.[10] A synthetic Y chromosome, in contrast, could be designed to just operate for one generation. Because it would not have to support evolution, it could have a much simpler structure. A disposable synthetic Y chromosome could be used to treat X-linked illness or infertility. In this case, it would be an advantage for the synthetic chromosome to be relieved from the requirements of evolution.

Evolution and Design

This example draws attention to an important issue for synthetic biology and for this book: the relationship between evolution and design. A key difference between them is that design is goal directed. Synthetic biology aims to design biological systems that perform predefined functions, whereas evolution has no directionality as it relies on random mutations. As was argued in previous chapters, intentional design necessarily incorporates values because judgments have to be made about what constitutes a good design and who the design is for. Such value judgments clearly do not play a role in evolution.

There are also differences between evolution and design in respect to control. The biologist François Jacob famously described evolution as "a tinkerer who uses everything at his disposal to produce some kind of workable object."[11] Evolution's tinkering may lead to beneficial and unexpected outcomes, but if evolution is used as a tool of synthetic biology, this will be accompanied by some loss of control. If we lose control, are we still designing?

Sheref would answer this question in the negative. When talking about the role of design in his research, Sheref opposed it to evolution. He said that although it is much easier for scientists to make biological systems evolve than purposely to design them, evolution is not design, precisely because of the loss of control. Nevertheless, evolution is still a desired endpoint of his protocell research, because, as we saw earlier, he considers evolution to be a defining feature of life. Sheref drew a distinction between designing something and letting it evolve, both of which he wants to achieve in his work. He talked of two phases, with phase 1 being design and phase 2 being evolution.

An opposing argument could be made, however, that evolution is itself a form of design. In directed evolution, in particular, it is possible to design the selection pressures precisely, so that evolution can only act within very narrow parameters, making it a powerful synthetic method.[12] If you select for precision, you will get precision. This idea of making use of evolution for our own design purposes is not new. In 1904, Hugo de Vries predicted that "evolution has to become an experimental science, which must first be controlled and studied, then conducted and finally shaped to the use of man."[13]

Returning to the point made earlier about the use of synthetic biology in increasing our understanding of biological systems, evolution can be a tool *for* synthetic biology, but we might also understand evolution better by using the methods *of* synthetic biology. This is because synthetic biology can be used as a starting point in an evolutionary process. Synthetic biologists can create novel permutations that are not found in existing organisms and see how evolution acts on them. Such work could provide a better understanding

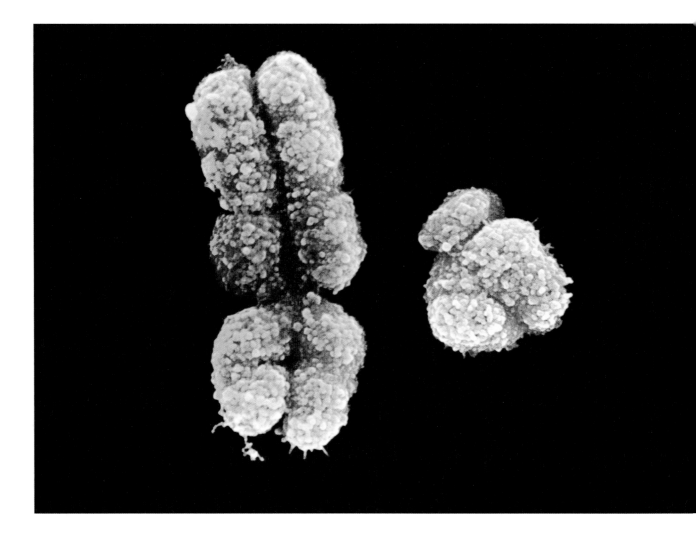

Figure 16.3
The human Y chromosome (right) does not have a matching pair. It is paired with the much bigger X chromosome (left).

of the mechanisms of evolution over the past 3 billion years. This understanding, in turn, could influence the way in which we design things—not only in synthetic biology but also in other fields.

In fact, evolution could even become an inspiration for engineering design. When synthetic biologists talk about making biology into something that can be engineered, they are usually thinking of existing engineering disciplines such as mechanical and electrical engineering. But rather than trying to force biology into the mold of these existing disciplines, perhaps what we should do is use evolution to change the way we engineer. Evolutionary computational algorithms mimic the evolutionary process to do design work (see chapters 1 and 7), and they are already used in the computing, automobile, and aerospace industries.[14] By learning more about how evolution works, we could see the rise of new engineering design paradigms that are based on evolution.

Neither Sheref nor Sascha are invested in engineering approaches to synthetic biology, and it is interesting that they used their Synthetic Aesthetics project to explore and challenge these approaches. In their joint work on living machines, they have raised questions about the extent to which existing engineering principles can be imposed on living things. This has opened up an important discussion about the relationship between evolution and design, which I have only scratched the surface of here. In their work, they also point to the idea that living machines are somewhat contradictory entities: For something to qualify as living it needs to evolve, but if it evolves, then our current ideas about reliable engineering may have to be rethought.

If art is about raising questions rather than providing answers, it is perhaps not surprising that bringing an artist together with a protocell scientist has given rise to a whole host of interesting questions about evolution and its role in the living machines of the future. These questions are important for synthetic biology because if it is going to succeed in designing living things, then it will be necessary to have ongoing conversations about the role of evolution in this process.

Cultures
and Context

17.
The Inside-Out Body

Christina Agapakis
and Sissel Tolaas

For in the eighteenth century there was nothing to hinder bacteria busy at decomposition, and so there was no human activity, either constructive or destructive, no manifestation of germinating or decaying life that was not accompanied by stench.

—Patrick Süskind, *Das Parfum*

Cheese is a product of human and bacterial collaboration, milk curdled and flavored with the by-products of microbial metabolism. The apocryphal origin story of cheese starts with milk stored in an animal skin thousands of years ago, the acids, enzymes, and microbes working to thicken the milk

Figure 17.1
Gorgonzola, Brie, Emmentaler, and Parmesan. Much of the diversity of appearance, taste, and smell is due to the varieties of species of bacteria and fungi that help make the cheese during the later stages of aging.

as well as fight off more harmful bacterial spoilage. Today, cheese production is a highly orchestrated and industrialized process of controlled rotting and thorough sterilization designed to ensure safety and consistency.

Raw milk contains an array of microbes that immediately begin to break down the emulsion of sugars, fats, and proteins. Modern cheese production starts with pasteurized milk seeded with pure strains of *Lactobacillus*—probiotic bacteria common in the human digestive tract and used in the production of fermented foods such as yogurt, sauerkraut, or kimchi. The lactobacilli ferment the sugar lactose, breaking it down into lactic acid as they grow and divide. As the acid builds up in the souring milk, the electrical charge on the emulsified milk proteins is neutralized. No longer charged and repelling each other, the proteins start to clump together into soft curds, separating from the liquid whey. Rennet, a cluster of stomach enzymes that helps young animals digest their mother's milk, is added to further break down the proteins and solidify the curds. Once solid, the curds can be drained, pressed, and processed into the huge variety of cheeses found around the world.

Milk from different animals can be curdled and processed using a range of techniques, but the true diversity in cheese flavor comes from the many species of bacteria and fungi that colonize the curds during the later stages of

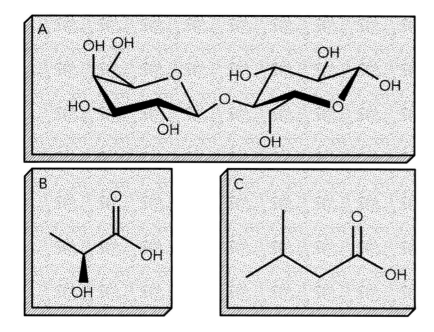

processing and aging. Fungi from the genus *Penicillium* produce both the stinky blueness of Gorgonzola and the mild white rind of Brie (figure 17.1). The flavor and the characteristic holes of Swiss Emmentaler cheese originate from the bacteria *Propionibacterium freudenreichii*, which consumes the lactic acid and produces other flavor molecules and carbon dioxide gas.

One of the main flavor compounds produced by *P. freudenreichii* is isovaleric acid, a molecule that has a pungent "cheesy" or "sweaty" smell. Indeed, the same molecule is found in human sweat and is the primary flavor note of body odor. This molecular link between cheese and the body reflects a biological link; the armpit is home to *Propionibacterium* species that break down oils from the skin through similar metabolic pathways to those involved in Swiss cheese production.

Likewise, the powerful foot-like odor of Limburger cheese is produced by *Brevibacterium linens,* a close relative of the *Brevibacterium epidermis* that lives between our toes. The smell of Limburger cheese is so strongly human that leaving a piece of it outside attracts the mosquitoes that preferentially bite at our feet and ankles.[1] It is not just mosquitoes that are fooled by the similarity of cheese and body odor. Cheese odors are associated with olfactory illusions—different perceptions of the same smell with different contextual cues. Telling someone that a pure mixture of isovaleric and butyric acid (another common cheese compound) is Parmesan cheese or vomit will lead to very different perceptions and very different emotional reactions (figure 17.2).[2]

Figure 17.2
Important molecules of cheese chemistry: (A) lactose, (B) lactic acid, and (C) isovaleric acid.

Isovaleric acid can smell deliciously cheesy in Swiss cheese but repulsively cheesy in body fluids. We categorize and compartmentalize smells, but more often opt out of smelling altogether, pushing for complete deodorization. This war on body odor began in the early twentieth century, with marketers convincing American women that they smelled bad.[3] Afraid of the humiliation of perceived dirtiness, we block and cover up odors on our bodies and in our homes, ignoring our noses and dulling our experience of the world. The push for deodorization extends even to cheese; as Camembert production has become increasingly industrialized and the microbes growing on each round cheese standardized, the flavors have been muted, the smells minimized, the flavor experience tempered.[4]

The fight against smells mirrors the fight against microbes, where the existence of a few deadly germs has led us to prefer complete sterilization to a more nuanced cohabitation with the many microorganisms that play a neutral or even positive role in human health. With the development of germ theory in the late nineteenth century began a campaign to eradicate the microbes that cause disease. Sterilization of surgical instruments, pasteurization of food products, and improvement of city hygiene saved countless people from dangerous infection, but caught in the crossfire were the millions of species of bacteria that are not just noninfectious, but actually required for the maintenance of environmental and human ecosystems. With the advent of metagenomic sequencing, the true diversity of microbial ecosystems is slowly becoming apparent. The importance of microbes to our digestive and immune health is leading to a renegotiation of our relationship with bacteria: no longer a wholesale destruction, but a careful coexistence and a nurturing of positive relationships. The social status of dirt is improving.

In a biological world characterized by enormous microbial diversity but a cultural world that still often emphasizes total antisepsis, cheese has also become a site of the growing post-Pasteurian debate of the importance of microbes in our lives.[5] Governments still regulate the sale of raw milk cheese, defining standards for pasteurization, aging, and testing. Slowly, the passion for probiotics is extending beyond the "slow food" community, exponentially increasing the number of live culture yogurt brands available in the supermarket. Microbes are making their way back into our diets, diversifying our gut flora, and changing the way that we think about the human body.[6]

Cheese making crafts flavors and odors through the management of microbial communities, connecting our work on odors and microbial engineering. Reproducing or creating new odors with chemistry requires a careful assessment of volatile molecules, a synthesis of chemicals from many different sources to create a coherent whole. Synthetic biologists likewise study, standardize, and catalog biological functions from many sources,

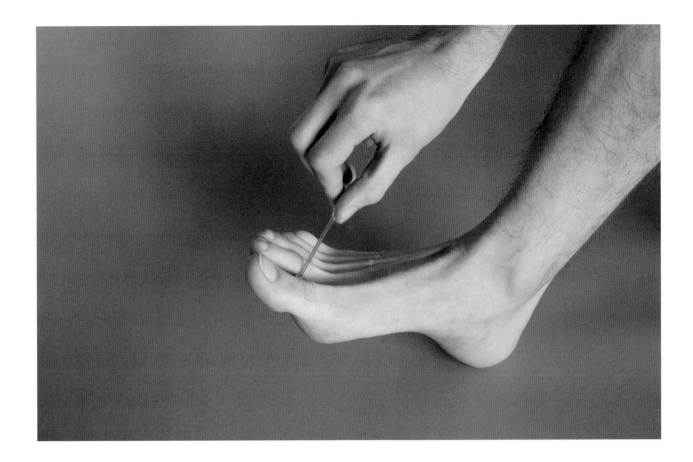

synthesizing and recombining them into new living systems. In making our own cheese for our work for the Synthetic Aesthetics project, we sought to highlight the processes of fermentation, the construction of microbial communities, the volatile molecules produced by living cells, and the connections between the microbial communities in cheese and those living on our skin.

We collected microbes from our own skin, rubbing cotton swabs over our hands and in between our toes, in our armpits, and inside our noses. Each swab was put into a small jar of organic pasteurized whole milk and warmed overnight. In the morning, the acids produced by the microbes had done their work. We could strain the curds away from the whey, making a series of small cheeses. These cheeses were of course not the aged masterpieces of artisan cheese makers, but microbial sketches, capturing some of the ecological diversity of different bodies and different body parts, bringing to the foreground the living odors of the body (figures 17.3–17.7).

Our final set of eight cheeses produced a wide range of smells, differing depending both on the person and on the body part that the swab had been taken from (table 17.1). Isolating individual bacterial colonies from each

Figure 17.3
Collecting bacteria from the human body to use to culture cheese.

Figure 17.4
Christina, left, and Sissel
preparing to make cheese from
human bacteria.

Figure 17.5
The cheese-making process
starts once the human bacteria
samples have been added. The
body-part bacterial samples were
used to make individual cheeses.

Figure 17.6
The curds of the "human cheese,"
strained after culturing.

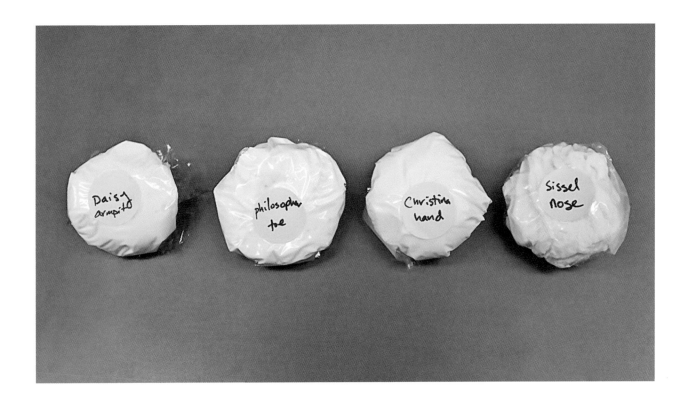

Figure 17.7
The final selection of human
body cheese: "Daisy Armpit,"
"Philosopher Toe," "Christina
Hand," and "Sissel Nose."

of the cheeses and using DNA sequencing to identify each species yielded
a surprisingly short list of strains, with striking commonality between the
different cheeses. The difference in the smells and the similarities in the species points to a deeper ecological diversity than that which can be isolated
using standard microbiological techniques.

Each species has its own unique smell in isolation (table 17.2), joining
in the chorus of volatiles from the milk and the other microbes to create
the full cheese odor. The overall odor can be isolated and analyzed using
techniques common to analytic chemistry and perfumery. The volatile
molecules in the headspace air around each cheese were captured using a
vacuum pump that pulls air through an absorbent silica matrix. The chemicals trapped in the matrix were then processed with gas chromatography–
mass spectrometry by International Flavors and Fragrances. The resulting
trace of peaks discloses the identity and relative quantity of each volatile
molecule (figure 17.8). In these traces, we found several cheese-associated
molecules, from the pleasing ketones of Armpit-3 to the isovaleric acid of
sweat and Swiss cheese in Hand-1 and Foot-5.

As the cheeses age, they pick up other microbial travelers, settling on
the slowly forming rind. These can be purposely seeded, as in industrial
cheese making, live in the cheese caves of artisanal cheese making, or attach

Table 17.1

Smell notes on cheeses produced from the microbes of different people and from different body parts

Source	Odors	Bacteria Isolated	Source	Odors	Bacteria Isolated
Hand-1	Yeast, ocean salt, sour old cheese, feet	*Providencia vermicola* *Morganella morganii* *Proteus mirabilis*	Armpit-2	Neutral, perfumed, industrial, synthetic, fermentation, car pollution, burning, sharp, chemical	*Enterococcus faecalis* *Hafnia alvei*
Foot-1	Sweat, big toe nail, cat feet, sweet, milky, orange juice in the fridge too long, fungus, buttery cheese, soapy, light perfume	*Providencia vermicola* *Morganella morganii* *Proteus mirabilis*	Armpit-3	Neutral, sour, floral, smooth, yogurt	*Microbacterium lactium* *Enterococcus faecalis* *Bacillus pumilus* *Bacillus clausii*
Armpit-1	Feta cheese, turkish shop, nutty, fruity, fishy	*Providencia vermicola* *Morganella morganii* *Proteus mirabilis*	Foot-5	Yeast, jam, feet, putrid, sour, rotten	*Providencia vermicola* *Proteus mirabilis*
Nose-2	Cheesy feet, cow, cheese factory, old subway station, toilet cleaner	*Providencia vermicola* *Morganella morganii* *Proteus mirabilis*	Armpit-4	Yogurt, sour, fresh cream, butter, whey	*Enterococcus faecalis*

Note: Volunteer "noses" include friends and colleagues, both artists and scientists, as well as the expert opinion of professional cheesemongers in a Berlin shop. Cheeses are identified by both the body part of origin, as well as the person that it came from (anonymized with a numerical suffix). Bacteria were isolated and sequenced from each cheese using standard microbiology techniques.

Table 17.2

Morphology and smell notes for the species identified in our cheeses

Bacteria	Appearance	Odors	Also Found
Providencia vermicola	Shiny white colonies	Sharp, vinegar, chlorine, swimming pool, sweet, floral, tulip	Gastrointestinal tract
Morganella morganii	Shiny yellow colonies	Pungent, rotting fish, dog breath, barn, monkey house at the zoo	Skin, airways, predatory ground beetle digestive tract, wallaby cloaca, frog skin, pea aphid, metal-working fluids and aerosols, histamine production in cheese
Proteus mirabilis/vulgaris	Fast-moving biofilm that creates bull's-eye appearance when spread over Petri dish.	Putrid, foul	Urogenital tract (can cause urinary tract infection), skin, airways, swine manure, cheese volatiles
Enterococcus faecalis	Small white colonies	Not much of a smell, chlorine, pool bathroom	Blood, diabetic wound microbiota, raw milk, cheese
Hafnia alvei	Dense and fluffy streaks	Sour, salty, corn tortillas, old leather couch, musty, gym mats	Gastrointestinal tract, human skin microbiome, feces of the pygmy loris, yellow catfish stomach, cauliflower flavor additive for cheese production
Microbacterium lactium	Bright yellow small opaque colonies	Hard crumbly cheese	Skin, Irish washed-rind cheese
Bacillus pumilus	Fluffy yellow	Deep-fried chicken, fried fat, cheddar cheese, Cheez-Its, Brie cheese	Soil, cheese spoilage
Bacillus clausii	White colonies	Stinky cheese, stinging, bleach, alcohol	Skin

Note: Smell notes provided by the amateur "noses" of scientists familiar with the smell of laboratory microbes.

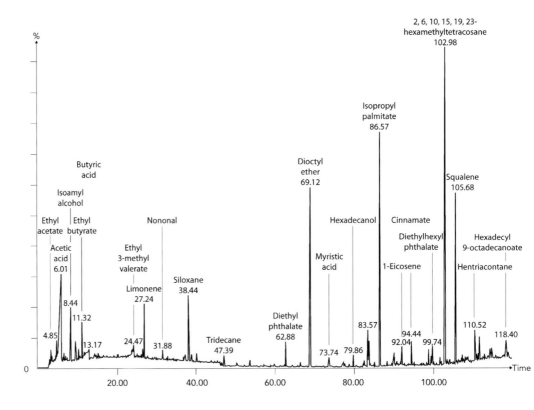

Figure 17.8
Gas chromatography–mass
spectrometry (GC-MS) trace for
the cheese "Nose-2." The timing
and size of peaks give identity
and relative concentration
information, respectively, of
the volatile molecules. GC-MS
performed by International
Flavors and Fragrances.

themselves accidentally in the refrigerator. These cheeses have grown and evolved in the years since we made them, ripening and molding, becoming organisms in their own right.

Our cheeses and our bodies are symbiotic complexes, home to many different species interacting to form a complete and evolving whole. Symbiosis is simply defined as "the living together of unlike organisms,"[7] a broad definition that covers parasitism, mutualism, and many relationships in between. The human body is a symbiotic collection of thousands of species, holding several pounds of microorganisms that outnumber human cells by a ratio of more than 10:1, just a small sample of which we were able to capture in our cheese.

Much of the diversity of the human skin microbiome is uncultivable, unable to grow in laboratory conditions or even in milk. We wanted to collect a more comprehensive library of those relatively few strains that are happy growing outside of the body, a collection of body microbes that we could draw from for future cheese making. Armed with Q-tips and test tubes, we held a "swabbing party" at a bar during the SXSW Interactive conference in Austin, Texas, capturing microbes from the crevices of 71 donors. From those samples, we isolated 30 unique strains, ranging in odor, color, and texture,

including several cheese-associated microbes such as *Staphylococcus sapro-phyticus* or *Microbacterium foliorum* (figure 17.9).

The list of all body microbes, however, also includes several strains that can cause disease when the ecosystem is out of balance: the staphylococci, pseudomonads, and streptococci of infection. In people with a compromised immune system or a weakened microbial ecosystem, infectious bacteria can overwhelm the community and take hold. To protect cheese from similar opportunistic infections that spoil the flavor and threaten the health of the consumer, the process of cheese making is primarily a process of cleaning—scrubbing and sterilizing pots and tools so that no undesirable bacteria take hold and ruin the rind ecosystem.

Our first batch of cheese didn't control for which species we collected from our skin and we thus cannot ensure its safety, leading to a "don't try this at home" warning. Isolating individual species or groups of species with no known role in infection and seeding them onto washed cheeses ready for aging will decrease the likelihood of food-borne illness in future batches, as will a deeper understanding and better ability to design healthy ecosystems.

Will we someday be able to improve our health by modulating the population of microbes we carry? Will we be able to design multispecies

Figure 17.9
Diversity of microbes isolated from human cheeses.

biotechnologies, strengthened by ecosystem bonds? Biotechnology and synthetic biology most often imagine bacteria or yeasts housed in isolation in shiny stainless-steel vats, but cells are rarely alone in nature. Organisms collaborate to do things no single strain could do alone, to digest complex molecules or to survive in harsh conditions. Designing for symbiosis rather than isolation opens up new frontiers for synthetic biology.[8] Perhaps rather than modeling synthetic biology on computer engineering, cheese making might be an engineering paradigm that allows for the design, construction, and maintenance of complex living worlds performing incredible feats of metabolism.

Cheese making, microbial ecosystems, and biotechnology each present examples of complex mixed cultures. All bring diverse groups of life forms together into intricate ecologies of competition and collaboration, affecting our culture and our environment. Heather Paxson, an anthropologist who studies the microbial politics of artisanal cheese making, writes on the interactions between human cultures and microbes, "To speak doubly of cheese cultures—bacterial and human—is thus no idle pun."[9]

As a cultural and biological object, cheese is an ideal "model organism" for the Synthetic Aesthetics project, likewise highlighting symbiosis at many scales. Rachel Dutton, a microbiologist at Harvard, dissects the rinds of artisanal cheeses, seeking to find in the symbiosis between the cheese fungi and bacteria a simplified system that can provide clues about how microbes work together in the much more complex environment of the skin.[10] The Synthetic Aesthetics project offers a model of how art and science can form lasting symbiotic relationships, an exploration into what is possible when unlike disciplines live together, new knowledge and new technology emerging at the intersection of multiple fields.

At the heart of synthetic biology is the often-uneasy relationship between biology and engineering, pursuing an understanding of life as it has evolved and life as it could be designed. These "two cultures," however, find themselves on the same side of the deeper rift between the arts and sciences, the seemingly irreconcilable split diagnosed so long ago by C.P. Snow.[11] But in that famous lecture, Snow also warned against drawing an easy line between any two: "The number 2 is a very dangerous number: that is why the dialectic is a dangerous process. Attempts to divide anything into two ought to be regarded with much suspicion." In cheese making, we see that the boundaries between human and microbe, art and science, and biology and engineering are fuzzy, mutating and evolving as we learn more about the value of ecosystems and complex mixed cultures.

18.
Transgressing Biological Boundaries

Alexandra Daisy Ginsberg

"The reason my cheese is so delicious," said Ms. Amieva, without a trace of modesty, "is my hands." She turned her meaty, callused palms over for inspection. "The natural bacteria in my skin makes the cheese more flavorful."[1]

This frank declaration from a venerable artisan cheese maker in remote northern Spain to a *New York Times* journalist may turn even hardy stomachs, but her observation brings into relief relatively recent prejudices about what we eat, our bodies, and the relationships between the two. I admit I was taken by surprise; I had never realized that cheese was still often made with bare hands dug into the pure, white milky curds, or that those same

hands vigorously scrub salt into the rinds of maturing cheese. With a modern preference for remotely processed, sanitized, quality-controlled, and comfortingly uniform food, the idea that someone else's naked, bacteria-rich skin should touch, let alone improve, food seems almost horrifying. Even stranger is the idea that eating cheese means eating something very much alive, which may even include living remnants of one's own body—or worse—of others' (figure 18.1).

Whether churned by whirring, steel robotic knuckles or tenderly crafted by hand following ancient methods, adding a culture of bacteria to milk as the start of a process toward preservation and portability is universal to cheese making around the world. And just as the 200 wheels of artisan cheese that Oliva Peláez Amieva makes each year are enhanced by a starter culture

Figure 18.1
Perhaps surprisingly, cheese makers often still stir the curds by hand. Their skin—and all its microbial residents, too—comes into contact with the food itself.

that includes her own microscopic helpers, Christina Agapakis and Sissel Tolaas's first batch of cheese was made with, and flavored by, bacteria. These bacteria, though, were harvested solely from our bodies, swabbed from the noses, toes, armpits, and crevices of those in Christina's lab one day.

Not long after our body cultures were first dipped into the fresh milk, the innocent-looking cheeses were passed around a packed wood-paneled lecture theater at MIT's Media Lab. Muffled squeals spilled from the room, incongruous against the quiet of the gleaming white-carpeted atrium outside, more used to dry technology of the silicon-based variety. The audience passed them along, tentatively smelling each cheese, some trying to distinguish themselves with a measured air of objectivity by comparing smelling notes with their neighbors. For others, the small, white, moist cheeses were just too much, transgressing accepted boundaries of decency, objects impossible to disassociate from their human origins. Each cheese emitted a personal odor, unique to its origins. Combined, the room was filled, quite literally, with a body of smells. "Daisy Armpit" had a fresh, yogurt-like aroma and was deemed the most mouth-watering, almost good enough to eat (the author writes, relieved). "Sissel Nose" and "Christina Mouth" were interesting and full-bodied in their own ways. "Philosopher Toe," however, was aggressively foul for all, a cheese whose footy origins were unassailable in its rancid bouquet. While the cheeses may be simple portraits of found organisms, with flavor profiles far less sophisticated than a long-perfected Reblochon or an expert Cheddar, the concept of human body cheese is powerful in itself.

Because these cheeses are not made from human fluids but from the bacteria living with us, calling them "human body cheeses" is a little misleading. Yet the remnant of instinctive disgust is curiously revealing about the relationship we have with the hidden bacteria on our bodies. If we do not like them, why do we still feel ownership or strong disgust when things from ourselves are disembodied, whether our bacteria into cheese or discarded hair or fluids? Can we disassociate them from their human origins? The experts in a Berlin cheese shop that Christina and Sissel visited a few weeks after the MIT session were not told of the origins of the cheeses. When they learned the truth after a session giving smelling notes on the samples, they were marvelously unperturbed.

For someone sheltered from the craft of cheese making, however, culturing food with materials harvested from the human body, from the private spaces of mouths, feet, and nooks to the open pastures of backs and necks and arms or hands with their different cultural meanings, challenges our relationship with our bodies, the organisms we share it with, and what we eat. Body cheese invites us to modify our assumptions and clear-cut classifications about the world. These cheeses may not be made using synthetic biology; instead, they use one of the oldest biotechnologies in the book. Crucially, they present a challenge to a future conflict between our antibacterial, pasteurized culture and the bacteria-enabled future that synthetic biology promises. All this must be critically examined.

You Are What You Eat

"You are what you eat!" is a familiar adage, whether an argument for the merit of the unfinished greens on your plate or an unsubtle admonishment against dipping back into a tasty pudding. The relationship between what we eat and our health is better understood than ever before: from the discovery of vitamins and programs of enrichment, to the distinction between good and bad fats, to finding links between diseases and foods, to the design of a balanced diet with five-a-day vegetable consumption as promoted by the U.K. government. We know, although we may choose to ignore it, that what we put into us has a direct impact on our bodies.

Food safety has gone hand-in-hand with this expansion of knowledge. The pasteurization of milk is now so commonplace that raw milk is effectively contraband in the United States, an unexpected hitch at the beginning of Christina and Sissel's cheese-making endeavor as they shopped for raw milk. The sterilization of cooking tools, long-life packaging, "best before" dates, and quality control all mean that for some of the world's population, we've never had cleaner, safer food or as much of it on our plates. Such "progress" appears inextricable from the industrialization of food (which paradoxically can be dirtier as animals are farmed at factory scale, requiring increased use of antibiotics) and a rejection of diversity. Straight carrots and identical chickens are symptoms of a decoupling between the reality of food production and how we imagine food as we experience it. Buying from supermarket shelves, we have become divorced from what we eat.

Science's discovery of the connection between bacteria and disease amid the nineteenth century hygiene movement, and the ensuing twentieth century modernist focus on clean urban living, have certainly contributed to extended life spans. But with awareness has come neurosis about the invisible threat of bacteria, to the point of societal intolerance of dirt. We know that the bacteria are there, unseen, multiplying, ready to strike and kill. Bacteria and the smell of unwashed human bodies that may harbor them are classified as dangerous. Bodies, pets, cars, houses are blitzed with antibacterials, sprayed with antiperspirants, or masked with fragrances and deodorants (figure 18.2). We scrub for fear of the uncivilized stink of the "natural," unclean, unwashed state that signifies dirt and potential disease. Household surfaces are liberally swabbed with products that boast of genocide on 99.99% of germs.

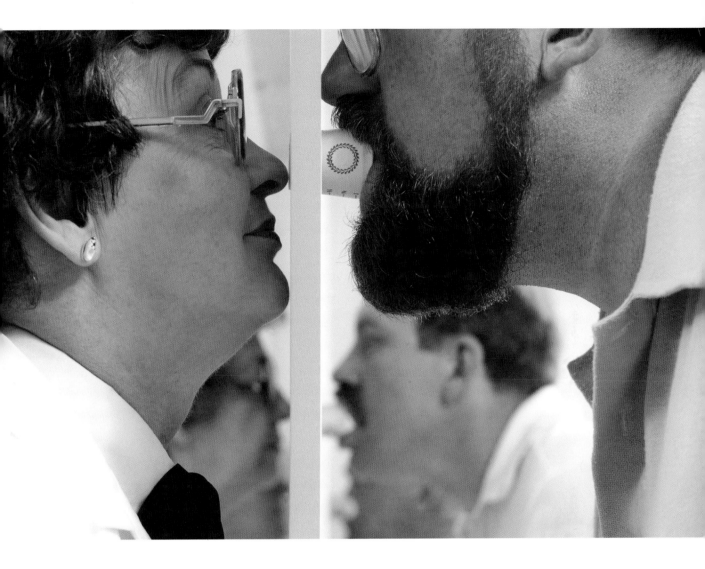

But while the aisles are stacked high and wide with antiseptic products, down in the dairy section, another world exists for the chosen few. Here, billions and billions of *Lactobacillus casei* are teeming in premium probiotic yogurts, promising to restore our guts to health (and originally sourced from someone else's gut). "Bacteria kill us but they can also save us?" we ask. This schizophrenic classification can only become more complex as we learn more about the benefits of the human microbiome, while advances in synthetic biology may present us with new kinds of consumer biological products to supplement the cheese, yogurt, and beer that we have savored for millennia. If we are so squeamish about bacteria, might synthetic biology need to rethink its promises of biotech consumer products, or should we change our thinking?

Figure 18.2
Researchers from Hill Top Research Inc. in Ohio smell the breath of research participants. The subjects' breath is disembodied and decontextualized by a temporary partition. Olfactory science demands that armpits, feet, and even genitals must be analyzed.

The Barriers between Biology and Us

There are four kinds of social pollution, according to anthropologist Mary Douglas: "The first is danger pressing on external boundaries; the second, danger from transgressing the internal lines of the system; the third, danger in the margins of the lines. The fourth is danger from internal contradiction, when some of the basic postulates are denied by other basic postulates, so that at certain points the system seems to be at war with itself."[2]

Concepts of social pollution can be biological in origin and even triggered by invisible or microbial forces. As you sit reading this book, you are absolutely not alone. Disregard any other people in the room; we are examining *you*, a person limited by the periphery of your body and your neat barrier of skin. We define ourselves as individuals, but this social construct is far from the physical reality. Teeming inside and on the surfaces of you are billions of invisible living things. While we like to ignore our biological state—the inescapable entropy of aging, our bodily emissions—the reality is that we are not just human, we are also biology. Your body is a menagerie, the bacteria you carry and support weighing in at as much as 3 to 6 pounds. Your personal biodiversity is as individual as your fingerprint, and liable to change in different stages in your life and state of health. Living clean of bacteria is impossible because they help you to survive (figure 18.3).

In rendering our human microbiome "smellable," tangible, and even edible, our symbiosis with bacteria becomes very real. We may not be able to see our bacteria, but imagine if we could see the body's boundaries as they are, loose and porous, as we travel through space, inhaling, exhaling, collecting new bacteria, shedding, strewing. Our physical borders are permeable; microbes infiltrate the environment and us, picking up DNA from other bacteria and leaving our own behind.

While we have learned to find bacteria disgusting—objects of horror as the invisible agents of disease—we cannot separate ourselves from "our own" bacteria. Harvesting and culturing the bacteria from the body into cheese reshapes this hidden part of our identity: these bacteria are an integral part of us. Human cheese brings our dirt into focus. We are no longer what we eat; rather, we eat what we are.

Is it just for this reason that Christina and Sissel's human cheese challenges us or are there other explanations as to why this project engenders such a strong gut reaction? Artists' experimentation with human body fluids is well-worn artistic territory. Piero Manzoni's 1961 "*Merda d'artista*," 90 cans of the artist's feces, priced at the weight of gold, set a precedent for a rich thread of effluent art, including menstrual blood, semen, sweat, and pus. Wim Delvoye's "*Cloaca*" is a series of elaborate disembodied mechanical digestive tracts, fed expensive dinners cooked up by chefs (figure 18.4).

Figure 18.3
The human microbiome is currently the subject of much study as researchers try to understand better the complex relationships we have with our microbial passengers.

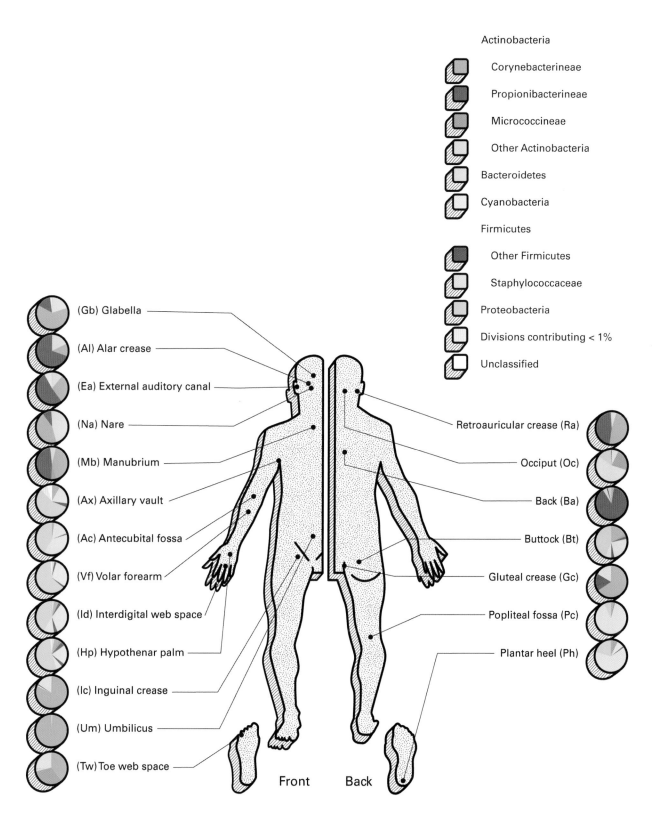

Actinobacteria
- Corynebacterineae
- Propionibacterineae
- Micrococcineae
- Other Actinobacteria

Bacteroidetes

Cyanobacteria

Firmicutes
- Other Firmicutes
- Staphylococcaceae

Proteobacteria

Divisions contributing < 1%

Unclassified

(Gb) Glabella
(Al) Alar crease
(Ea) External auditory canal
(Na) Nare
(Mb) Manubrium
(Ax) Axillary vault
(Ac) Antecubital fossa
(Vf) Volar forearm
(Id) Interdigital web space
(Hp) Hypothenar palm
(Ic) Inguinal crease
(Um) Umbilicus
(Tw) Toe web space

Retroauricular crease (Ra)
Occiput (Oc)
Back (Ba)
Buttock (Bt)
Gluteal crease (Gc)
Popliteal fossa (Pc)
Plantar heel (Ph)

Front   Back

As the food is processed into perfectly formed turds, they shatter the pristine gallery context that they occupy. More recently, we have seen cheese made from the breast milk of the New York designer Miriam Simun, more breast milk cheese served up by New York chef Daniel Angerer at his Klee Brasserie, and in London, £14 *Baby Gaga* ice cream made from breast milk being banned, all generating enormous publicity from our repulsion. But why do we have this aversion to stuff from our bodies, and do different areas of the body engender different reactions of distaste?

How would you react if you walked into a room to find someone putting their shoes on the table? You may be outraged enough to reprimand them for their unreasonably dirty behavior. But we are normally perfectly happy to touch shoes, keeping them so close to our bodies that we carry them around all day. Shoes are not intolerably dirty to touch in themselves. But when their relative context is changed, moved from the floor to the table, their modified place in the accepted pattern of things renders them much dirtier. Mary Douglas uses this example in *Purity and Danger*, a book that reappraised how we see dirt and our cultural attitudes to it.[3]

Cheese smells are reminiscent of what we have learned to classify as bad odors on our own bodies. These aromas, once taken out of context and served up on a plate, become irresistible to the cheese lover, who may even think, "the stinkier the better." But body cheese confuses the accepted patterns of the world around us. Things that lived invisibly and ignored on our body are now visible and made dirty. According to our cultural patterns, human bacteria should not be in cheese. Human cheese is a symbol out of place, dangerous as it transgresses accepted social boundaries.

While work in the human microbiome shows that certain parts of the body are "dirtier" than others, in that they harbor more bacteria than others, they are still clean in context, as some of those bacteria are working with our body to fight off infection. Meanwhile, "cleaner" bits of the body, those with fewer bacteria, may be concealing far more dangerous residents. Are some areas more or less suitable for making food? Could vagina cheese be less dirty than elbow cheese, a far more culturally neutral body part? Even those words sound prohibitively disgusting, placed together. Mouths, sheltering all manner of bacteria, have vastly conflicting symbolic associations, both erotic and foul. Kissing and spitting may involve the same fluids, but context is all-important. Watching Christina sample her own mouth cheese, carried with her in a Tupperware container on the plane all the way from Los Angeles to Austin, Texas, for an iteration of the project, was gag-inducing for me as an observer. As she put the cheese down after taking a bite, Christina smiled, remarking that it tasted pleasant.

Perhaps it is the associations we attribute to different areas of the body or the idea that the sweat is toxic, as it is a waste product pushed out. The armpits and groin are full of apocrine sweat glands that pump out milky sweat that contains proteins and possibly pheromones, which are hungrily digested by bacteria; it is the by-products of this process that make it smell more strongly. Is this why sweaty armpit cheese is more disgusting than shoulder cheese, where local eccrine glands simply help with cooling, producing water and salts? There are negligible amounts of sweat on a tiny swab, in comparison to the billions of bacteria that will multiply to help coagulate the cheese, but we find it hard to relinquish the connection between waste and contaminated food. Douglas suggests why we see effluence from the body (feces, mucus, blood, vomit, urine) or material discarded (skin, hair, nails) as disgusting. Matter discharged from the body is marginal; it reminds us that the structure, the body itself, is vulnerable at margins. Once things leave our control, their status is unclear. Human body cheese carries this lack of clarity with it, its existence dependent on human refuse.

Bodies carry symbolism in almost all cultures, as models of society. Cultures that perform sacrifices are well aware of this, enacting rites on the body of an animal whose power will carry across to their groups as a whole. These rituals of body symbolism cause an emotive individual experience, and human cheese is no different.[4] It is in itself a microcosm, a model of the body it came from and the cultures on it. Human cheese threatens our concept of the body's solid boundaries and threatens community boundaries as it breaches established cultural norms of sanitation and etiquette.

This is a foodstuff that carries a latent suspicion of cannibalism. The cultural rule that food goes into the body and does not come from it is disrupted. Would eating cheese harvested from your own body be more problematic, a sensation of eating a model of your own body, or is it worse to eat others'? Christina explains that people were much more open to eating their own cheese or smelling their own bacteria than smelling someone else's. "After people got over the disgust of thinking about their own bacteria, they also felt ownership over it. This is 'mine' and this is 'me.'" Might different cheeses from different body parts and people carry different meanings? If body cheeses were developed beyond individual body portraits to more complex ecosystems, engineered hybrids of different people or organisms, might they stretch further boundaries?

Mary Douglas compares tribal initiation rites that include defilement of different sorts as an essential way for societies to thrive: just as in gardening, composting is essential as the old, the dead, and waste matter is folded back in to enable growth. We have the ability to design new patterns, symbols, and classifications. Human body cheese shocks us out of our malaise, making us consider differently the raw material of biotechnology: biology. This focus includes bacteria, our own body, and the symbiosis between different biological scales and communities. Body cheese helps us to experience biology anew. As we learn more, how might we challenge these boundaries and relationships?

Growing a Biological Future

When it comes to synthetic biology, its practitioners are planning to engineer things that are inherently part of us all. As Christina and Sissel's human body cheese shows us, we are much closer to the bacteria that we are planning to engineer than abstract circuit diagrams relay. Through the shock of a tangible artifact, *The Inside-Out Body* draws biology out of the conceptual discussions of synthetic biology. Faced with the individuality of living things, does the uniform engineering vision, eliminating diversity in favor of a controllable uniformity, seem as desirable?

How might we rethink our relationships with nature through biology and our understanding of it? There are many ways that we might work symbiotically with the natural living world. Body cheese gives us new ways of thinking about synthetic biology beyond industrial fermenters. The promised biological future may be one of bacteria pumping out fuels in hidden vats, but it also hints at new biotechnologies in which the body becomes a site or a starting culture for manufacture. With the growing focus on the human microbiome as an undeniable part of ourselves, our attitudes toward helpful bacteria may grow ever friendlier as we understand how much we rely on them: Research shows that bacteria in our gut may even impact what we choose to eat. Proposals to "decontaminate" astronauts of existing bacteria for long missions, replacing their personal colonies with novel, controlled species, may be more than a safety precaution to remove parasites that feed off our waste. Engineering the human microbiome may not be so far removed from ethical debate over engineering the human body.

Understanding these cheeses as models of the body, the rendering of parts of each person's body in an external object, hints at new kinds of products. Already, Rachel Dutton at Harvard University is looking at cheese as a form of model organism, to explore complex relationships in ecosystems. Human-originating cheeses could offer study of individuals' body flora, with a view to future products far more aligned to our bacterial populations, whether personalized consumer probiotics, or inspiring new approaches to medicines that avoid rejection by the body, grown in or outside of the body, or finding ways to avoid the incompatibility of synthetic materials within the human body, such as transplants or implants.

Human cheese reminds us that we are biological entities; just like the bacteria synthetic biologists are engineering. We are part of nature, not separate from it. Operating as individuals neither works in biology nor in culture. Designing objects and cultures, biological or not, that function in a vacuum is impossible in reality. It can even be damaging to think of the things we make as isolated from context, as we miss interacting with the whole. Synthetic biology wants to teach biology how to behave, but biology has much to teach us, even about our own bodies, identities, and cultural assumptions.

Epilogue:
Reflections on Synthetic Aesthetics

When we started this project, we did not expect that it would involve discussions of futility in the dust of the Western Australian desert, the creation of a circular sound installation from a bacterial plasmid, heated arguments over the differences between "nature," "biology," and "life," or the sampling of smells from human cheeses. The project has taken us to labs, studios, galleries, seminars, deserts, and even NASA. It has brought together many voices, often with very different, sometimes even contradictory, things to say. The diversity of views found in this book does not lead to a single conclusion; instead, it shows the breadth of different ways of exploring synthetic biology and design. But there are themes that emerge from this diversity, and we focus on three of these here: the unique nature of biology, the inseparable connection between design and values, and the difficulties of occupying the future. We end by asking how this project should be classified and what it has produced: is it art, design, synthetic biology, or something else altogether?

The Nature of Biology

Biology evolves. This is not a surprising statement, but it becomes important when we start thinking about design, because design does not normally work with things that evolve. This point is central to an exploration of synthetic biology and design, and it is addressed in almost all of the chapters in this book. As Sascha and Sheref put it, once we start designing things that evolve, we will be "designing with Darwin," and we will have to change our understanding of design (chapter 15). Because biology evolves, it exhibits "fabulous willfulness and malleability" (chapter 10), which presents considerable challenges not only to design but also to engineering. The arguments in this book suggest that engineering in biology may look rather different from engineering in other areas.

Several of the residencies either implicitly or explicitly demonstrated the limitations of engineering metaphors when applied to biology. For example, David and Fernan became very interested in the way in which biology solves spatial problems—that is, how it "computes"—through shape and form (chapter 7). This is in sharp contrast with how we normally think of biological "computation," through an analogy with digital computation. In converting synthetic biological data into sound, Chris and Mariana also revealed the limitations of the types of representations used in engineering approaches to synthetic biology (chapter 13). Because the pair had to make decisions about which aspects of the data they should represent as sound, they drew attention to what normally gets left out in standard visual representations of synthetic biology, which typically depict "parts" and "devices" in "circuit diagrams." Sissel and Christina were explicit about the limitations of computer engineering metaphors in synthetic biology and suggested an alternative: cheese making (chapter 17). This, they argued, has advantages over abstract circuit diagrams because it allows us to think in terms of complex and dynamic ecosystems.

The contributions to this book have not only shown the limitations of engineering metaphors but also have drawn attention to the unique properties of biology. The idea of "harnessing the complexity of biology in a manner sympathetic to its capabilities" (chapter 10) is something that was a theme of many of the residences. For example, an advantage of biology is that it can be used in a more sustainable way than other materials, and building materials that are made through bacterial manufacturing are likely to be cheaper, with better performance and fewer carbon emissions (chapter 7). Such materials are also more likely to be available locally, reducing the pressure on transport infrastructure.

One potentially beneficial property of biology that several of the residencies highlighted is its sensitivity to its surrounding environment. Sascha and Sheref explored the idea that the "living machines" created by synthetic biology would be able to evolve along with their environment in a manner that would be more sustainable in the longer term than the machines we use today (chapter 15). Wendell and Will speculated about synthetic biological products that might have beneficial relationships not only with their broader environments but also with their human users—such as the team's personalized skin-care product (chapter 9). Sissel and Christina went a step further by suggesting that such developments in synthetic biology might even change our relationship to the natural world. Rather than thinking of bacteria as something that should be eliminated, instead we may come to value "a careful coexistence and a nurturing of positive relationships" (chapter 17). All these examples show that one of the great potentials of synthetic biology is to exploit the ability of biology to coexist with us symbiotically.

But these unique properties of biology are often overlooked in discussions of the industrialization of synthetic biology. Instead, we are presented with a vision of "drop-in" replacements, which use bacteria as a mechanism to produce more of what we already have, rather than doing something more interesting that draws on the particular characteristics of biological systems (chapter 1). Such invisible replacements perpetuate unsustainable manufacturing practices because they do not require that we change our behavior. As Daisy puts it in chapter 3, what we see is "a disruptive technology that also promises to disrupt nothing." Such applications of synthetic biology are focused on the near future. They are an attempt to demonstrate that synthetic biology can slot in with our current lifestyles and demands.

As well as not paying adequate attention to the nature of biology, the industrial framing does not allow for playfulness and creativity, which are essential for imaginative innovation. Chapter 10 is a plea to retain a place for creativity in synthetic biology, provoked by the collaboration between product designers and synthetic biologists in chapter 9. This collaboration showed that science and engineering were actually just as messy and creative as design, and just as likely to involve confusion and inspiration. Creativity, humor, and play were themes of several of the other residencies. Sissel and Christina's cheese project is the one that is most likely to produce laughter in an audience, but this laughter is not superficial—it is also an expression of mild discomfort. Something that is necessary for play is the freedom to pursue work that is seemingly useless. This was a theme of Oron and Hideo's collaboration (chapter 11), which directly challenged the short-term emphasis on applications that we see in much synthetic biology and pushed us to consider the value of artistic and scientific work that has no obvious utility. Exploration and play as ends in themselves are already features of synthetic biology, exhibited most clearly by the iGEM competition (see chapters 2 and 6) and the "Do It Yourself" biology movement. The contributions to this book show that play is something that should be taken seriously.

Design and Values: Making Synthetic Biology "Better"

Synthetic biology models itself on existing design practices, but, like current models of industrialization, this book has shown that these design practices are themselves unsustainable. Design has become a service for facilitating mass production, instead of a way of making things meaningful. As Daisy argued in chapter 3, we must seize the opportunity to reinvent design as we reinvent biology. This could take the form of critically engaged biological design.

Critically engaged biological design would embrace the inescapably "value-laden character of design."[1] Design involves values because when something is designed, this necessarily raises a whole host of further questions such as: Is it designed well or not? For what purpose is it designed? And who is it designed for?[2] Importantly design takes us from "what is?" to "what could be?" Design is about possibility. It necessarily projects itself into the future—a future that is imagined to be better than the present.

This ties into one of the original objectives of the Synthetic Aesthetics project: to make synthetic biology better. What is meant by "better" is clearly the key question here. Does it mean technically better; that is, cheaper, faster, easier, and more predictable? Does it mean more useful or more profitable? More sympathetic to the nature of biological systems? More responsible or ethical? More sustainable, more beautiful, or more democratic? Many of these ways of making synthetic biology better were exhibited in the residencies, showing that the question requires us to think not only about what defines the parameters of good biological design but also about who makes those judgments. The involvement of artists and designers provides new entry points into these discussions.

Recognizing the importance of values in design and engineering also provides novel ways to engage with ethical issues. For example, once we start thinking about design processes, rather than asking questions like "should we play God?" we can ask "What intentions and assumptions are driving this practice of design?" (chapter 5). The latter question is much more tangible than the former, it is clearer how to answer it, and it has more direct relevance to the daily activities of scientists and engineers. Another way of engaging with ethical questions is by taking a historical perspective, as demonstrated by Oron and Hideo's and Sascha and Sheref's contributions (chapters 11 and 15, respectively). Both these chapters show that our current designs are imbued with values that we may not previously have acknowledged (about the nature of machines, for example). Oron and Hideo's temporal perspective also raises questions about our responsibility to the future. It encourages us to reconsider our own place in the geological timescale and to show humility about all human activities. These examples show how the residencies have raised important ethical issues, but in new ways.

Occupying the Future: How to Inhabit the Spaces of Promise?

Design takes us into the future by imagining the possible. Synthetic biology is also positioned in the future because it is, by and large, a technology that does not yet exist. As previous chapters have shown, there are few synthetic biology applications as yet, and the field is primarily driven by its potential. For these reasons, this book is itself helping to create the future of the field. By focusing on synthetic biology, it is making synthetic biology more real than it would be otherwise. This construction of the future of synthetic biology was built into the project from the start, as the "sandpit" that funded the project was specifically focused on synthetic biology (see the Introduction chapter). In fact, it is because synthetic biology is a new field that it was open to the kinds of interdisciplinary interactions initiated by the Synthetic Aesthetics project. All the residents selected to participate had to engage with the topic of synthetic biology, and in doing this almost all of them imagined a future where synthetic biology was a successful technology. It is easy to forget that many of the outputs of the residencies are fictional. Will and Wendell's packaging that builds its own contents (chapter 9) is a computer-manipulated image, as are many of the images on these pages. The book is in this sense a work of fiction, but arguably no more so than many of the policy narratives about the potential of the field.

When working in such a future-oriented area, a key question that arises is how to inhabit the spaces of promise? The residencies described in this book demonstrate six different ways of answering this question. They show that it is possible to occupy the future by bringing art and design into the early stages of a technology and by speculating about the directions this technology might take. Although Wendell and Will were the pair that most explicitly initiated an open discussion about the future of synthetic biology, this was a theme of all the residencies (as well as the *E. chromi* and *Growth Assembly* work described in chapter 6). In engaging with the future, none of the residents repeat familiar narratives of dystopia (synthetic biology destroys the world) or utopia (synthetic biology saves the world), but instead force us to consider the possibility of living with designed biological things.

Art and design are not the only way of inhabiting the spaces of promise of emerging technologies. The field of science and technology studies (STS) has developed methods such as "constructive technology assessment"[3] and "anticipatory governance,"[4] where social scientists work with scientists and engineers to broaden the design of a technology to integrate social, ethical, political, and economic considerations. These methods recognize the central role of values in design and use tools such as scenarios, fictions, and foresight to help make these values explicit, so that different—perhaps conflicting—visions of the future can be compared and debated. What the Synthetic Aesthetics project has shown is that many artists and designers and scholars in STS share the aim of having an "enlarged conversation"[5] about synthetic biology, being involved in discussions across disciplinary and institutional boundaries, and engaging with different ways of imagining the future.

How to Classify the Project?

In the context of these cross-boundary discussions, it is much harder to divide people into "insiders" and "outsiders," and something that is particularly interesting reflecting on the Synthetic Aesthetics project is that the artists and designers involved resist easy classification.

One way in which it might be superficially attractive to classify their work is as an example of "Do It Yourself" biology (DIYbio).[6] This is a loosely defined heterogeneous mixture of people who share an interest in "doing biology outside of traditional professional settings."[7] Although artists and designers have a history of being involved in non-institutional biology,[8] the Synthetic Aesthetics project is dissimilar from DIYbio in several ways. The scientific work in the project took place in established research institutions, so it cannot be described as being non-institutional science. The project was also funded by national science funding agencies (which incidentally raises bigger questions about the obligations that participants have, explicitly or implicitly, toward their funders, especially in a politicized field). However, by doing things like making cheese in a laboratory or growing algae on a circuit board, the residents did subvert the tools of synthetic biology for the purposes of art or design. They had different aims and motivations from those that are normally found in scientific institutions, so in this way their work has some similarities with DIYbio.

Although we were clear from the start that the outcome of this project was not science communication, another way in which it might be tempting to classify the exchanges is as a form of public engagement with science. The artists and designers, after all, were members of the public who were allowed into science laboratories for a limited period of time. But in many ways they were not at all like the "public" (itself an extremely problematic term, as STS has shown). The residents learned scientific tools and techniques, including the assembly of BioBricks (David), lipid vesicle construction (Sascha), and cyanobacterial protocols (Oron). They got their hands "wet" and were not simply observing the laboratory from afar. Their involvement meant that they became implicated in the biotechnological transformation of living things. Perhaps more importantly, in the other half of the exchange, expertise and skills were passed from the artist/designer to the scientist/engineer. For example, Fernan learned about computational optimization of building design from David, and Sissel taught Christina about the practice of smell design. In these situations, the scientists and engineers were the lay people in the exchange.

What Have We Produced?

However we choose to categorize the residents, we set them a hard task. They had to engage with recalcitrant biological materials, with a propensity to evolve, which do not fit neatly into an engineering framework but which do have unique properties. They also had to negotiate the relationship between design and values and operate in the unstable and unknown space of the future. In dealing with these challenges, what have they produced? Is it art? Is it science? Is it synthetic biology? Is it social science? Is it something different altogether?

To focus on one example, the cheeses produced by Sissel and Christina are simultaneously scientific and artistic objects. Scientifically, by creating cheese with a tangible form and a pungent smell, the pair has provided accessibility to the otherwise invisible bacterial microbiome that surrounds us. Artistically, their project raises profound questions about the boundaries between our bodies and the food that we eat. Most importantly, a common reaction to their project is: "I need to go and think about that."

Like the human cheeses, what has come out of the Synthetic Aesthetics project and this book cannot be easily classified. Rather than communicating knowledge about synthetic biology to "the public" or helping synthetic biology contribute to the economy by facilitating industrialization, what we are seeing in the work presented here is "the production of new objects and practices of knowledge."[9] Such new objects and practices will perhaps be the most important outputs of the Synthetic Aesthetics project, and these may in the future contribute to challenging and transforming the way we think about art, design, synthetic biology, and society.

By rethinking existing disciplines in this way, we hope to help prevent synthetic biology simply following unimaginative and entrenched paths. Standing in the way of closure by provoking debate and encouraging discussion has become our key goal in writing this book. Rather than offering easy answers, we have raised new questions about design and synthetic biology. We hope that we have also inspired our readers to ask new questions of their own.

Notes

Introduction

1. In 2011, Joy Y. Zhang, Claire Marris, and Nikolas Rose cited 39 reports since 2004 alone. See Joy Y. Zhang, Claire Marris, and Nikolas Rose, "The Transnational Governance of Synthetic Biology: Scientific Uncertainty, Cross-borderness and the 'Art' of Governance," *BIOS Working Paper* No. 4 (London: London School of Economics and Political Science, 2011).

Chapter 1

1. Chris French et al., "Development of biosensors for the detection of arsenic in drinking water," in *Biosensors*, ed. J. Santini (London: Elsevier, 2012).

2. James D. Watson and Francis H. C. Crick, "A structure for deoxyribose nucleic acid," *Nature* 171, no. 4356 (1953): 737–738.

3. Francis Crick, "Central dogma of molecular biology," *Nature* 227 (August 1970): 561–563.

4. Barry Canton, Anna Labno, and Drew Endy, "Refinement and standardization of synthetic biological parts and devices," *Nature Biotechnology* 26, no. 7 (July 2008): 787–793.

5. Richard S. Kirby, *Engineering in History* (Mineola, NY: Courier Dover Publications, 1990), 171.

6. Jason R. Kelly, Adam J. Rubin, Joseph H. Davis, Caroline M. Ajo-Franklin, John Cumbers, Michael J. Czar, et al., "Measuring the activity of BioBrick promoters using an in vivo reference standard," *Journal of Biological Engineering* 3, no. 4 (March 2009).

7. Jessica S. Dymond, Sarah M. Richardson, Candice E. Coombes, Timothy Babatz, Héloïse Muller, Narayana Annaluru, et al., "Synthetic chromosome arms function in yeast and generate phenotypic diversity by design," *Nature* 477 (September 2011): 471–476.

8. Jason Chin, T. Ashton Cropp, J Christopher Anderson, Mridul Mukherki, Zhiwen Zhang, and Peter J. Schultz, "An expanded eukaryotic genetic code," *Science* 301, no. 5635 (August 2003): 964–967.

9. Wenlin An and Jason Chin, "Synthesis of orthogonal transcription–translation networks," *Proceedings of the National Academy of Sciences* 106, no. 21 (May 26, 2009): 8477–8482.

Chapter 2

1. H.G. Wells, "The limits of individual plasticity," in *H.G. Wells: Early Writing in Science and Science Fiction*, ed. Robert Philmus and David Y. Hughes (Berkeley: University of California Press, 1975): 36.

2. Synthetic Biology, "Synthetic biology: FAQ." Available at http://syntheticbiology.org/FAQ.html.

3. Daniel G. Gibson, John I. Glass, Carole Lartigue, Vladimir N. Noskov, Ray-Yuan Chuang, Mikkel A. Algire et al., "Creation of a bacterial cell controlled by a chemically synthesized genome," *Science* 329, no. 5987 (July 2010): 52–56.

4. Andrew Pollack, "Custom-made microbes, at your service," *New York Times* (January 17, 2006).

5. Wells, "The limits of individual plasticity," 36.

6. Jacques Loeb, "On the nature of the process of fertilization and the artificial production of normal larvae (plutei) from the unfertilized eggs of the sea urchin," *American Journal of Physiology* 3 (1899): 135–138; reprinted in *Studies in General Physiology* (Chicago: University of Chicago Press, 1905): 539–543.

7. Jacques Loeb to Ernst Mach, 28 December 1899, EM; on the myth of the hero in late nineteenth-century German science see Frank J. Sulloway, *Freud, Biologist of the Mind* (New York: Basic Books, 1979): 445–495. Quoted in Scott F. Gilbert, *Developmental Biology* (New York: University Press, 2000): 93–117.

8. Philip J. Pauly, *Controlling Life: Jacques Loeb and the Engineering Ideal in Biology* (Oxford: Oxford University Press, 1987): 47.

9. Pauly, *Controlling Life,* 47.

10. William Bechtel and Robert C. Richardson, "Vitalism," in *Routledge Encyclopedia of Philosophy*, ed. E. Craig (London: Routledge 1998). Vitalism is the idea that "living organisms are fundamentally different from non-living entities because they contain some non-physical element or are governed by different principles than are inanimate things."

11. Jacques Loeb, *The Dynamics of Living Matter* (New York: Columbia University Press, 1906), 223.

12. Alexis Carrel, *Man, the Unknown* (New York: Harper & Bros, 1935).

13. Carrel, *Man, the Unknown*, 299.

14. Gordon Rattray Taylor, *The Biological Time Bomb* (New York: The New American Library, 1968).

15. TED, "Craig Venter unveils 'synthetic life'." Available at http://www.ted.com/talks/craig_venter_unveils_synthetic_life.html.

16. Joe Davis, "Microvenus," Art Journal 55, no. 1 (Spring 1996): 70–74.

17. ArtScienceBangalore, "iGEM team wiki '09." Available at http://2009.igem.org/Team:ArtScienceBangalore (for information about the Srishti team project in iGEM).

18. The Cambridge 2009 iGEM team had Alexandra Daisy Ginsberg and James King, both graduates from the Design Interactions Department at the Royal College of Art, as their advisors.

19. For example, Joseph Straw, "Securing Synthetic Biology," *Security Management*. Available at http://www.securitymanagement.com/article/securing-synthetic-biology-008829?page=0%2C1.

20. Diego Gambetta and Steffen Hertog, "Why are there so many engineers among Islamic radicals?" *European Journal of Sociology* 50, no. 2 (2009): 201–230.

21. Gambetta and Hertog, "Why are there so many engineers among Islamic radicals?" 201–230.

22. Emmanuel Sivan, "Why are so many would-be terrorists engineers?" *Haaretz* (February 12, 2010).

23. Jonathan Safran Foer, *Eating Animals* (London: Penguin, 2009): 34.

24. New Harvest, "Advancing alternatives to conventionally produced meat through research and education." Available at http://www.new-harvest.org/.

25. Winston Churchill, "50 years hence," *Popular Mechanic* (March 1932).

26. SymbioticA—Centre of Excellence in Biological Arts. Available at http://www.symbiotica.uwa.edu.au.

Chapter 3

1. Victor Margolin and Richard Buchanan, eds., *The Idea of Design* (Cambridge, Mass.: MIT Press, 1995): xiv.

2. *The Oxford English Dictionary* (Oxford: Oxford University Press, 2012).

3. Bruno Latour, "A cautious Prometheus? A few steps toward a philosophy of design (with special attention to Peter Sloterdijk)," Keynote speech, *Networks of Design from Design History Society* (Cornwall, September 3, 2008): 5. Available at http://www.bruno-latour.fr/sites/default/files/112-DESIGN-CORNWALL-GB.pdf.

4. Jean Baudrillard, *The System of Objects* (London: Verso, 1996): 146 [originally published by Editions Gallimard, 1968].

5. Deyan Sudjic, *The Language of Things* (London: Allen Lane, 2008).

6. Latour, "A cautious Prometheus," 2.

7. Mat Hunter, "What design is and why it matters," Design Council. Available at http://www.designcouncil.org.uk/about-design/What-design-is-and-why-it-matters/.

8. John Thackara, *In the Bubble: Designing in a Complex World* (Cambridge, Mass: MIT Press, 2005): 196.

9. John Rousseau, "Design after design," *Design Mind*, no. 16 (April, 2012).

10. Bruce Mau, "Massive change is not about the world of design, it's about the design of the world." Quoted in Léa Gauthier, *Design for Change* (Paris: Blackjack éditions, 2011): 22.

11. Bruno Latour, "Love your monsters," in *Love Your Monsters: Postenvironmentalism and the Anthropocene,* ed. Michael Shellenberger and Ted Nordhaus (Oakland, Calif.: Breakthrough Institute, 2011).

12. Christien Meindertsma, Katherine Rosmalen, and Joel Gethin Lewis, *PIG 05049,* (Rotterdam: Flocks, 2007).

13. European Commission, "GM Food & Feed Labelling." Available at http://ec.europa.eu/food/food/biotechnology/gmfood/labelling_en.htm

14. George Gessert, *Green Light toward an Art of Evolution* (Cambridge, Mass.: MIT Press, 2010): 8.

15. Food and Agriculture Organization of the United Nations, "What is Agrobiodiversity?" 2004. Available at ftp://ftp.fao.org/docrep/fao/007/y5609e/y5609e00.pdf.

16. Richard Pell, Center for PostNatural History. Available at http://www.postnatural.org/.

17. Perhaps synthetic biology should still aspire to "design for biology" or "with biology" instead of aiming for the "design of biology." See Meredith Walsh, "Design for Life," ISEA2011 Istanbul. Available at http://isea2011.sabanciuniv.edu/paper/design-life.

18. Gessert, *Green Light toward an Art of Evolution,* 14.

19. Latour, "A cautious Prometheus?" 5.

20. Nicholas Wade, "Researchers say they created a 'synthetic cell,'" *New York Times* A17 (May 21, 2010).

21. James F. Danielli, "Industry, society and genetic engineering," *The Hastings Center Report* 2, no. 6 (December 1972): 5–7.

22. Graham Chedd, "Danielli the Prophet." *New Scientist and Science Journal* (January 21, 1971): 124–125.

23. Presidential Commission for the Study of Bioethical Issues, *New Directions: The Ethics of Synthetic Biology and Emerging Technologies* (Washington, D.C.: U.S. Government Printing Office, 2010).

24. John Wyndham, *The Day of the Triffids* (London: Penguin Books, 1951).

25. Frederik Pohl and Cyril M. Kornbluth, *The Space Merchants* (New York: Ballantine Books, 1953).

26. John Christopher, *The Death of Grass* (London: Michael Joseph, 1956).

27. Robert Macfarlane, "Life in ruins," Darwin Lecture from University of Cambridge (Cambridge, England, January 27, 2012). Available at http://www.sms.csx.cam.ac.uk/media/1210608.

28. Brian Wilson Aldiss, *Billion Year Spree: The True History of Science Fiction* (Garden City, N.Y.: Doubleday, 1973).

29. Paul Rabinow and Gaymon Bennett, *Designing Human Practices: An Experiment with Synthetic Biology* (Chicago: University of Chicago Press, 2012).

30. Tom Bieling, Interview with Paola Antonelli, "Thinking and problem making," *Design Research Network* (February 17, 2009). Available at https://www.designresearchnetwork. org/drn/content/interview-paola-antonelli%3A-%2526quot%3Bthinking-and-problem-making%2526quot%3B.

31. Paola Antonelli, "Core Principles," February 10, 2009, *SEED Magazine*. Available at http://seedmagazine.com/content/article/core_principles/.

32. Glen Lowry, "Foreword," in *Design and the Elastic Mind*, ed. Paola Antonelli (New York: Museum of Modern Art, 2008): 4.

33. Anthony Dunne and Fiona Raby, Interviewed by Bruce Tharp of Core77 Broadcasts (December 17, 2007). Recording at http://www.core77.com/blog/broadcasts/core77_broadcasts_anthony_dunne_fiona_raby_interviewed_by_bruce_tharp_8433.asp.

34. David Crowley, "Design at the end of the world," *Icon Magazine* 064 (October 2008): 202.

35. Bieling, Interview with Paola Antonelli.

36. R. Buckminster Fuller, *Ideas and Integrities* (New York: Collier Books, 1969 [originally published 1963]): 176.

37. Natalie Jerimijenko, "One Trees: An Information Environment," 2003. Available at http://www.nyu.edu/projects/xdesign/onetrees/.

38. Ed Regis, *What is Life? Investigating the Nature of Life in the Age of Synthetic Biology* (New York: Farrar, Straus & Giroux, 2008): 7.

39. Frank Furedi, *Where Have All the Intellectuals Gone? Confronting 21st Century Philistinism* (New York: Continuum, 2004): 24.

40. Arthur Fine, "Fictionalism," *Midwest Studies In Philosophy* 18, no. 1 (May 1993): 1–18.

41. Macfarlane, "Life in ruins."

42. Hans Vaihinger and Charles K. Ogden, *The Philosophy of As If, A System of the Theoretical, Practical and Religious Fictions of Mankind* (London: K. Paul, Trench, Trubner & Co., 1924).

43. Dunne & Raby, "What If…," 2009. Available at http://www.dunneandraby.co.uk/content/bydandr/496/0.

44. Philips Design Probes, "Microbial Home," 2011. Available at http://www.design.philips.com/about/design/designportfolio/design_futures/microbial_home.page.

45. Eduardo Kac, *Signs of Life—Bio Art and Beyond* (Cambridge, Mass.: MIT Press, 2007): 18.

Chapter 4

1. Rob Carlson, "Planning for Toy Story and Synthetic Biology: It's All About Competition (Updated)." Available at http://www.synthesis.cc/2013/04/updated-dna-cost-and-productivity-curves-plus-a-few-more-thoughts-on-moores-law.html.

2. Harald Brüssow, Carlos Canchaya and Wolf-Dietrich Hardt, "Phages and the evolution of bacterial pathogens: from genomic rearrangements to lysogenic conversion," *Microbiology and Molecular Biology Reviews* 68, no. 3 (September 2004): 560–602.

3. Frederick Sanger, Gillian M. Air, Bart G. Barrell, Nigel L. Brown, Alan R. Coulson, J. C. Fiddes, et al., "Nucleotide sequence of bacteriophage φX174 DNA," *Nature* 265, no. 5596 (1977): 687–95.

4. Francis Crick and Leslie Orgel, "Directed panspermia," *Icarus* 19 (1973): 341.

5. William Bialek, Rob de Ruyter van Steveninck, and collaborators, "Physical principles in neural coding and computation: Lessons from the fly visual system" (May 2, 2007). Available at: http://www.princeton.edu/~wbialek/flypapers_links.html.

6. Steven Benner, "Understanding nucleic acids using synthetic chemistry," *Accounts of Chemical Research* 37, no. 10 (October 2004): 784–97.

7. Jerome Bonnet, Pakpoom Subsoontorn, and Drew Endy, "Rewritable digital data storage in live cells via engineered control of recombination directionality," *Proceedings of the National Academies of Sciences of the United States of America* 109, no. 23 (2012): 8884–9.

8. Jerome Bonnet, Peter Yin, Monica E. Ortiz, Pakpoom Subsoontorn, and Drew Endy "Amplifying genetic logic gates," *Science* 340, no. 6132 (May 3, 2013): 599–603

9. Monica Ortiz and Drew Endy, "Engineered cell-cell communication via DNA messaging," *SJournal of Biological Engineering* 6, no. 16 (September 7, 2012) doi:10.1186/1754-1611-6-16

Chapter 5

1. Richard Pierson and Alexander Bozmoski, "Harley-Davidson's 100th anniversary: The sound of a legend," *Sound and Vibration* (March 2003): 14–17.

2. Michael B. Sapherstein, "The trademark registrability of the Harley-Davidson roar: A multimedia analysis," *Boston College Intellectual Property and Technology Forum* (1998). Available at http://bciptf.org/1998/10/11/sapherstein/.

3. Associated Press, "Harley-Davidson seeks trademark on engine sound" (March 25, 1996).

4. Richard H. Lyon, "Product sound quality—from perception to design," *Sound and Vibration* (2003). Available at http://www.sandv.com/downloads/0303lyon.pdf.

5. Sharon Schembri, "Reconstructing brand experience: The experiential meaning of Harley-Davidson," *Journal of Business Research* 62, no. 12 (December 2009): 1299–1310.

6. Pablo Schyfter, "Technological biology? Things and kinds in synthetic biology," *Biology & Philosophy* 27, no. 1 (January 2012): 29–48.

7. For example, see chapter 1 in this book.

8. Nina Pope, *Cat Fancy Club*. Available at http://www.di.research.rca.ac.uk/content/projects/205/Cat+Fancy+Club.

9. Celia Hartmann, "Edward Steichen archive: Delphiniums blue (and white and pink, too)" (March 8, 2011). Available at http://www.moma.org/explore/inside_out/2011/03/08/edward-steichen-archive-delphiniums-blue-and-white-and-pink-too.

10. For example, see Pamela Silver, "Making biology easier to engineer," *Biosocieties* 4, no. 2–3 (September 2009): 283–289; or Drew Endy, "Foundations for engineering biology," *Nature* 438, no. 7067 (November 2005): 449–453.

11. Adam Arkin, "Setting the standard in synthetic biology," *Nature Biotechnology* 26, no. 7 (2008): 771–774.

12. Drew Endy and Adam Arkin, "A standard parts list for biological circuitry," Defense Advanced Research Projects Agency white paper (Berkeley, CA: 2009). Available at http://hdl. handle.net/1721.1/29794.

13. Barry Canton, Anna Labno, and Drew Endy, "Refinement and standardization of synthetic biological parts and devices," *Nature Biotechnology* 26, no. 7 (July 2008): 787–793.

14. Drew Endy, "Foundations for engineering biology," *Nature* 438, no. 7067 (November 2005): 449–453.

Chapter 6

1. David Attenborough, director, "Episode 2: Growing." *The private life of plants.* [documentary]. (BBC TV/Turner Home Entertainment, 1995). Poisonous Pitcher Plant sequence available at http://www.youtube.com/watch?v=trWzDlRvv1M.

2. Richard Fleischer, director, *Soylent Green* [film] (Metro-Goldwyn-Mayer, 1973).

3. Imperial College iGEM 2008 team, "Biofabricator Subtilis: Designer Genes." Available at http://2008.igem.org/Team:Imperial_College.

4. Alexandra Daisy Ginsberg and Sascha Pohflepp, "Growth Assembly," 2009. Available at www.daisyginsberg.com/projects/growthassembly.html.

5. Haeckel, Ernst. *Kunstformen der Natur* (Leipzig und Wien: Verlag des Bibliographischen Instituts, 1904).

6. The White House, *National Bioeconomy Blueprint* (Washington, D.C.: The White House, 2012). Available at http://www.whitehouse.gov/sites/default/files/microsites/ostp/national_bioeconomy_blueprint_april_2012.pdf.

7. Drew Endy, "Foundations for engineering biology," *Nature* 438, no. 7067 (November 24, 2005): 449–453.

8. Oxitec Ltd., "RIDL Science." Available at: http://www.oxitec.com/ridl-science/.

9. Peter Rüegg, "Sperm out of a capsule," ETH Life (May 2, 2011). Available at http://www.ethlife.ethz.ch/archive_articles/110501_Besamung_per/index_EN.

10. Ignacio García-Bocanegra, Rafael Jesús Astorga, Sebastián Nappa, Jordi Casal, Belén Huerta, Carmen Borge, and Antonio Arenas, "Myxomatosis in wild rabbit: Design of control programs in Mediterranean ecosystems," *Preventive Veterinary Medicine* 93, no. 1 (January 1, 2010): 42–50.

11. Bernard London, *Ending the Depression Through Planned Obsolescence* (New York: Author, 1932). Available at http://www.adbusters.org/blogs/blackspot_blog/consumer_society_made_break.html.

12. London, *Ending the Depression Through Planned Obsolescence.*

13. Jane Spencer, "Companies slash warranties, rendering gadgets disposable," *Wall Street Journal* (July 16, 2002). Available at http://online.wsj.com/article/0,,SB1026764790637362400,00.html.

14. Arup Group, "Sustainable to Evolvable," ed. Rachel Armstrong, 2012. Available at http://thoughts.arup.com/post/theme/6.

15. Monsanto's pledge has so far held but may be reignited due to an ongoing seed-patent case underway in 2013 in the U.S. Supreme Court, Vernon Hugh Bowman v. Monsanto Co. Open Letter from Monsanto CEO Robert B. Shapiro to Rockefeller Foundation President Gordon Conway and Others (October 4, 1999). Available at http://www.monsanto.com/newsviews/Pages/monsanto-ceo-to-rockefeller-foundation-president-gordon-conway-terminator-technology.aspx.

16. Alexandra Daisy Ginsberg and Oron Catts, "The Well-Oiled Machine," *Icon Magazine*, The Fiction Issue (Issue 80), February 2010.

17. William McDonough and Michael Braungart, *Cradle to Cradle: Remaking the Way We Make Things* (New York: North Point Press, 2002).

18. Malcolm Moore. "Tourists flock to Beijing's Olympic stadium even as dust gathers." *The Telegraph* (London, UK), August 7, 2009. Available at http://www.telegraph.co.uk/sport/olympics/5989713/Tourists-flock-to-Beijings-Olympic-stadium-even-as-dust-gathers.html.

19. Nicholas Spencer, "Wooden ambition," *Financial Times* (November 13, 2009). Available at http://www.ft.com/cms/s/0/c7612b28-cf1a-11de-8a4b-00144feabdc0.html.

20. Suzanne Lee, "BioCouture." Available at http://biocouture.co.uk/.

21. Stefan Schwabe, "The Kernels of Chimaera," 2012. Available at http://www.di12.rca.ac.uk/projects/the-kernels-of-chimaera/.

22. Ecovative Design, "Mushroom packaging," Available at http://www.mushroompackaging.com/.

23. Jae Rhim Lee, "Infinity Burial Project." Available at http://infinityburialproject.com/burial-suit.

24. Damian Arnold, "Self-healing concrete," *Ingenia* 46 (March 2011): 39–43. Available at http://www.ingenia.org.uk/ingenia/articles.aspx?index=654.

25. Richard Lenski, "E. coli long-term experimental evolution project." Available at http://myxo.css.msu.edu/ecoli/ /.

26. Jerome Bonnet, Pakpoom Subsoontorn, and Drew Endy, "Rewritable digital data storage in live cells via engineered control of recombination directionality," *Proceedings of the National Academy of Sciences of the United States of America,* published online before print (May 21, 2012): doi:10.1073/pnas.1202344109.

27. David Perlman, "Plan to merge labs for biofuel research criticized," SFGate (March 30, 2012). Available at http://www.sfgate.com/science/article/Plan-to-merge-labs-for-biofuel-research-criticized-3442863.php.

28. Synthetic Biology Project, "Risk assessment prior to field release," (2010–2013). Available at http://www.synbioproject.org/scorecards/recommendations/risk-assessment/risk-assessment-prior-to-field-release/.

29. Emma Frow, David Ingram, Wayne Powell, Deryck Steer, Johannes Vogel, and Steven Yearley, "The politics of plants," *Food Security* 1, no. 1 (February 2009): 17–23.

30. Suzanne LaBarre, "The Better Brick: 2010 Next Generation winner," *Metropolis Mag.Com* (May 12, 2012). Available at http://www.metropolismag.com/May-2010/The-Better-Brick-2010-Next-Generation-Winner/.

31. University of Cambridge iGEM 2009 team, "E. chromi." Available at http://2009.igem.org/Team:Cambridge.

32. Alexandra Daisy Ginsberg and James King, with the University of Cambridge 2009 iGEM team. *E. chromi.* Available at http://daisyginsberg.com/projects/echromi.html.

33. Harvard University iGEM 2010 team, "Human Practices." Available at http://2010.igem.org/Team:Harvard/human_practices/debate.

34. BCL (Shiho Fukuhara and Georg Tremmel), "Common Flowers/Flower Commons," 2008. Available at http://www.common-flowers.org/.

35. J. Paul Neeley, "Gaia Corporation," 2011. Available at http://www.jpaulneeley.com/gaiacorporation.html.

36. The BioBrick Foundation, "The BioBrick™ Public Agreement (BPA)." Available at https://biobricks.org/bpa/.

37. Paola Antonelli, Adam Bly, Lucas Dietrich, Joseph Grima, Dan Hill, John Habraken, et al., "Open Source Architecture (OSArc)," *Domus* 948 (May 11, 2011). Available at http://www.domusweb.it/en/op-ed/open-source-architecture-osarc-/.

38. Global Alliance for the Rights of Nature, "Rights of nature articles in Ecuador's Constitution," 2008. Available at http://therightsofnature.org/wp-content/uploads/pdfs/Rights-for-Nature-Articles-in-Ecuadors-Constitution.pdf.

39. I admit an interest in this project, having spent two weeks working with the team, running a workshop with James King to develop ideas. ArtScienceBangalore iGEM 2010 team, "Synthetic/Post-Natural Ecologies." Available at http://2010.igem.org/Team:ArtScienceBangalore/Project.

40. ArtScienceBangalore iGEM 2011 team, "Searching for the ubiquitous genetically engineered machine." Available at http://artscienceblr.org/index.php?/project/searching-for-the/.

41. University of Edinburgh iGEM 2006 team, "Arsenic Biosensor." Available at http://2006.igem.org/wiki/index.php/University_of_Edinburgh_2006.

42. This echoes an ongoing debate in the design world about "design imperialism." See Bruce Nussbaum, "Is Humanitarian Design the New Imperialism?" Fast Co Design (July 10, 2010). Available at http://www.fastcodesign.com/1661859/is-humanitarian-design-the-new-imperialism.

43. Penny Bailey, "The biggest poisoning in history," Wellcome Trust (January 24, 2013). Available at http://www.wellcome.ac.uk/News/2013/Features/WTP051176.htm

44. Paul Rabinow and Gaymon Bennett, *Designing Human Practices: An Experiment with Synthetic Biology* (Chicago: University of Chicago Press, 2012), 5-6.

45. Andrew Barry, Georgina Born, and Gisa Weszkalnys, "Logics of interdisciplinarity," *Economy and Society* 37, no. 1 (February 2008): 20–49.

Chapter 7

1. Werner E. G. Müller, Xiaohong Wang, Fu-Zhai Cui, Klaus P. Jochum, Wolfgang Tremel, Joachim Bill, et al., "Sponge spicules as blueprints for the biofabrication of inorganic–organic composites and biomaterials," *Applied Microbiology and Biotechnology* 83, no. 3 (June 2009): 397–413.

2. Sangok Seok, Cagdas D. Onal, Kyu-Jin Cho, Robert J. Wood, Daniela Rus, and Sangbae Kim, "Meshworm: A peristaltic soft robot with antagonistic nickel titanium coil actuators," *IEEE/ASME Transactions on Mechanotronics* PP, no. 99 (2012): 1–13.

3. Saket Navlakha and Ziv Bar-Joseph, "Algorithms in nature: The convergence of systems biology and computational thinking," *Molecular Systems Biology* 7 (November 2011): 546-556.

4. Yehuda Afek, Noga Alon, Omer Barad, Eran Hornstein, Naama Barkai, and Ziv Bar-Joseph, "A biological solution to a fundamental distributed computing problem," *Science* 331, no. 6014 (January 2011): 183–185..

5. Masatoshi Yamaguchi, Nadia Goué, Hisako Igarashi, Misato Ohtani, Yoshimi Nakano, Jennifer C. Mortimer, et al., "Vascular-related NAC-DOMAIN6 and vascular-related NAC-DOMAIN7 effectively induce transdifferentiation into xylem vessel elements under control of an induction system," *Plant Physiology* 153, no. 3 (May 2010): 906–914. This involved the use of *Arabidopsis thaliana* transgenic lines 35::VND6-GR and 35::VND7-GR and dexamethasone application.

6. For this experiment, we prepared protoplasts (isolated cells with no cell wall) from VND7 lines and used microfluidic devices to grow and trap them.

7. In both of these approaches, all of the xylem patterns were extracted using confocal microscopy and reconstructed in silico for analysis. High-resolution images were used to analyze the behavior of the xylem program in response to a large variety of envelope shapes and to gain more insights into this patterning process..

8. Michael Schmidt and Hod Lipson, "Distilling free-form natural laws from experimental data," *Science* 324, no. 5923 (April 2009): 81–85. This software application by Hod Lipson and Michael Schmidt at the Computational Synthesis Lab at Cornell University allows a user to launch a genetic algorithm to evolve and fit equations to match the data.

9. We supposed that each of the original bars of the exoskeleton could be one of three types (thick, thin, or deleted), and we solved for two fitness criteria: least structural displacement and least amount of material.

10. See Ginger Krieg Dosier's "Biomanufactured Brick." Available at http://www.sesam-uae.com/sustainable-business/profiles/ginger-krieg-dosier.asp.

11. See Suzanne Lee's "BioCouture." Available at http://www.biocouture.co.uk.

12. First, a customized interface of a 3D modeling software application (Autodesk Maya) allows users to adjust inputs for a bounding box, a chemical environment, and a tissue scaffold of nonbiological material. Second, the same interface allows users to adjust inputs for two or more types of bacteria, including their concentration, their starting location, their emission of a chemical, their attraction to a chemical, the material properties of the compounds they deposit, and ordinary differential equations to describe other more complex

behaviors. Third, this interface simulates the self-organizing behavior of the bacteria inside the environment. Self-organization refers to the emergence of organization from a homogeneous mix of components without external intervention and purely due to local interactions among the parts. Next, we applied multiobjective optimization routines. Each design permutation of a bacterial pattern and its corresponding deposition of two or more types of material can be tested for its mechanical performance in a multimaterial finite element analysis package (this is a customized version of VoxCAD, developed by Jonathan Hiller and Hod Lipson from the Computational Synthesis Lab at Cornell University). The inputs (properties of environment and bacteria) and outputs (mechanical performance) were used in a process of evolutionary computing and multiobjective optimization.

13. Andrew Adamatzky and Jeff Jones, "Road planning with slime mould: If Physarum built motorways it would route M6/M74 through Newcastle," *International Journal of Bifurcation and Chaos* 20, no. 10 (October 2010): 3065–3084. For instance, the use of foraging strategies by slime molds to achieve optimum network connectivity in maps is a good example of exploiting biological computation.

14. Peter Fratzl, "Biomimetic materials research: What can we really learn from nature's structural materials?" *Journal of the Royal Society Interface* 4, no. 15 (August 2007): 637–642.

15. Atsushi Tero, Seiji Takagi, Tetsu Saigusa, Kentaro Ito, Dan P. Bebber, Mark D. Fricker, et al., "Rules for biologically inspired adaptive network design," *Science* 327, no. 5964 (2010): 439–442.

Chapter 8

1. Keller Rinaudo, Leonidas Bleris, Rohan Maddamsetti, Sairam Subramanian, Ron Weiss, and Yaakov Benenson, "A universal RNAi-based logic evaluator that operates in mammalian cells," *Nature Biotechnology* 25, no. 7 (May 2007): 795–801.

2. Timothy S. Gardner, Charles R. Cantor, and James J. Collins, "Construction of a genetic toggle switch in *Escherichia coli*," *Nature* 403, no. 6767 (January 2000): 339–342.

3. Michael Elowitz and Stanislas Leibler, "A synthetic oscillatory network of transcriptional regulators," *Nature* 403, no. 6767 (January 2000): 335–338.

4. Caroline M. Ajo-Franklin, David A. Drubin, Julian A. Eskin, Elaine P.S. Gee, Dirk Landgraf, Ira Phillips, and Pamela A. Silver, "Rational design of memory in eukaryotic cells," *Genes & Development* 21, no. 18 (September 2007): 2271–2276.

5. Atsushi Tero, Seiji Takagi, Tetsu Saigusa, Kentaro Ito, Dan P. Bebber, Mark D. Fricker, et al., "Rules for biologically inspired adaptive network design," *Science* 327, no. 5964 (January 2010): 439–442.

6. See, for example, George Lakoff, *Metaphors We Live By* (Chicago: University of Chicago Press, 1981).

Chapter 9

1. Design at IDEO has roots in defining the look and feel of physical and interactive products for diverse industries such as telephony, computing, and health care and now includes defining the look and feel of services and spaces. Design involves a combination of processes and strategies for the future that will impact broadly people's experience of the world around them.

2. François Jacob, *Of Flies, Mice, and Men,* translated by Giselle Weiss (Cambridge, Mass.: Harvard University Press, 1998).

3. U. Alon, "How to choose a good scientific problem," *Molecular Cell* 35, no. 6 (September 2009): 726–728.

4. Some of the publications in which our work has been featured are an article in the journal *Cell* ("Drop the pipette: Science by design"), a nonscientific poster at the Synthetic Biology 5.0 conference at Stanford (2011), and an article exploring the potential applications of synthetic forms of life in the technology and design publication *Fast Company*.

Chapter 10

1. François Jacob, *The Statue Within: An Autobiography* (London: Unwin Hyman, 1988).

2. Uri Alon, "How to choose a good scientific problem," *Molecular Cell* 35, no. 6 (September 2009): 726–728.

3. Charles P. Snow, *The Two Cultures and The Scientific Revolution* (Cambridge, England: Cambridge University Press, 1960).

4. Plato, "Ion," in *Plato: The Collected Dialogues*, ed. Edith Hamilton and Huntington Cairns (Princeton, N.J.: Princeton University Press, 1961): 215–228.

5. Henri Poincare, "Mathematical creation," in *The Creative Process: Reflection on Invention in the Arts and Sciences*, ed. Brewster Ghiselin (Berkeley: University of California Press, 1980): 22–31. Reprinted from *The Foundations of Science*, trans G. B. Halstead (London: Science Press, 1924): 383–394.

6. Michael Davis, "An historical preface to engineering ethics," *Science and Engineering Ethics* 1, no. 1 (1995): 33–48.

7. Edward de Bono, *Six Thinking Hats* (New York: Harper & Row, 1985).

8. Edward de Bono, *Lateral Thinking: Creativity Step by Step* (New York: Harper & Row 1970).

9. Stan Liebowitz and Stephen E. Margolis, "The fable of the keys," *Journal of Law and Economics* 33, no. 1 (April 1990): 1–26.

Chapter 11

1. Jan Zalasiewicz, Mark Williams, Alan Smith, Tiffany L. Barry, Angela L. Coe, Paul R. Brown, et al., "Are we now living in the Anthropocene?" *GSA Today* 18, no. 2 (February 2008): 4–8.

2. Zalasiewicz et al., "Are we now living in the Anthropocene?"

3. Aurora Algae, "Growing natural solutions." Available at http://www.aurorainc.com/.

4. Erik Davis, "DeLanda Destratified." Available at http://www.techgnosis.com/delandad.html.

Chapter 12

1. Erik Davis, "DeLanda destratified." Available at http://www.techgnosis.com/delandad.html.

2. Raeid M. M. Abed, Sergey Dobretsov, and Kumar Sudesh, "Applications of cyanobacteria in biotechnology," *Journal of Applied Microbiology* 106, no. 1 (2009): 1–12.

3. Strategy for the UK Life Sciences, December 5, 2011. Available at https://www.gov.uk/government/uploads/system/uploads/attachment_data/file/32457/11-1429-strategy-for-uk-life-sciences.pdf, p. 10.

4. Christopher J. Preston, "Synthetic biology: Drawing a line in Darwin's sand," *Environmental Values* 17, no. 1 (2008): 23–39.

5. Steven A. Benner, Slim O. Sassi, and Eric A. Gaucher, "Molecular paleoscience: Systems biology from the past," *Advances in Enzymology and Related Areas of Molecular Biology* 75 (2005): 1–132.

6. SymbioticA. "Biological arts." Available at http://www.symbiotica.uwa.edu.au/.

7. Christopher M. Kelty, "Outlaws, hackers, victorian amateurs: Diagnosing public participation in the life sciences today," *Journal of Science Communication* 9, no. 1 (March 2010): C03.

8. Herbert A. Simon, *The Sciences of the Artificial*, second edition (Cambridge, Mass.: MIT Press, 1981): 55.

9. Harold G. Nelson and Erik Stolterman, *The Design Way* (Englewood Cliffs, N.J.: Educational Technology Publications, 2003).

10. Adrian Madlener, "What will it mean to be a designer in the future?" in *New Realities, New Roles for Designers*, Essay Competition (Eindhoven: Design Academy Eindhoven, 2011).

11. Klaus Ottmann, "Interview with Richard Serra," *Journal of Contemporary Art* (1989). Available at http://www.jca-online.com/serra.html.

12. Antony Dunne, *Herzian Tales* (Cambridge, Mass.: MIT Press, 1999).

13. Paola Antonelli, "Design and being just," *Seed Magazine* (March 23, 2009). Available at http://seedmagazine.com/content/print/of_design_and_being_just/.

14. Anthony Dunne and Fiona Raby, "Manifesto #39 Dunne & Raby," *Icon* 050, no. 23 (August 2007). Available at http://www.iconeye.com/news/manifestos/manifesto-39-dunne-amp-raby-%7C-designer#39-dunne-amp-raby-|-designer.

15. Anthony Dunne and Fiona Raby, *a/b, A Manifesto*. Available at http://www.dunneandraby.co.uk/content/projects/476/0.

Chapter 13

1. Vesa Välimäki, Julian D. Parker, Lauri Savioja, Julius O. Smith, and Jonathan S. Abel, "Fifty years of artificial reverberation," *IEEE Transactions on Audio, Speech, and Language Processing* 20, no. 5 (July 2012): 1421–1448.

Chapter 14

1. John Dewey, *Art as Experience* (New York: Perigee, 2005 [originally published 1934]): 86.

2. Dewey, *Art as Experience*, 230.

3. Dewey, *Art as Experience*, 230.

4. Drew Endy, "Foundations for engineering biology," *Nature* 438, no. 7067 (November 2005): 449–453.

5. Trevor Pinch, "Towards an analysis of scientific observation: The externality and evidential significance of observational reports in physics," *Social Studies of Science* 15, no. 1 (February 1985): 3–36.

6. Hiroko Terasawa, Josef Parvizi, and Chris Chafe, "Sonifying ECoG seizure data with overtone mapping," *Proceedings of the 18th International Conference on Auditory Display* (June 2012). Available at http://hdl.handle.net/1853/44445.

Chapter 15

1. Clay McShane and Joel Tarr, "The Decline of the Urban Horse in American Cities," *The Journal of Transport History* 24, no. 2 (2003): 177–198.

2. McShane and Tarr, "The Decline of the Urban Horse in American Cities."

3. Carl Warren Gay, *Productive Horse Husbandry* (Philadelphia: J.B. Lippincott, 1914).

4. Gay, *Productive Horse Husbandry*.

5. Daniel Margocsy, "Designing the horse: From Albrecht Dürers studies of proportions to the science of breeding in Stuart England," talk given at "Prints and the Pursuit of Knowledge in Early Modern Europe" Symposium, Harvard University, Cambridge, Mass., December 2–3, 2011.

6. Lily E. Kay, *Who Wrote the Book of Life?: A History of the Genetic Code* (Stanford: Stanford University Press, 2000).

7. Bruno Latour, *Pandora's Hope: Essays on the Reality of Science Studies* (Cambridge, Mass.: Harvard University Press, 1999).

8. Richard E. Lensky, "E. coli Long-term Experimental Evolution Project." Available at http://myxo.css.msu.edu/ecoli.

9. John W. Haas, Jr., "The Reverend Dr William Henry Dallinger, F.R.S. (1839–1909)," *Notes and Records of the Royal Society London* 54, no. 1 (January 2000): 53–65.

Chapter 16

1. Sheref S. Mansy, "Model protocells from single-chain lipids," *International Journal of Molecular Sciences* 10 (2009): 835–843.

2. Daniel G. Gibson, John I. Glass, Carole Lartigue, Vladimir N. Noskov, Ray-Yuan Chuang, Mikkel A. Algire, et al. "Creation of a bacterial cell controlled by a chemically synthesized genome," *Science* 329, no. 5987 (July 2, 2010): 52–56.

3. Steven A. Benner, "Q&A: Life, synthetic biology and risk," *BMC Biology* 8 no. 1 (2010): 77.

4. See also chapter 6.

5. Jeffrey Mervis, "Digging for fresh ideas in the sandpit," *Science* 324, no. 5931 (May 29, 2009): 1128–1129.

6. Drew Endy, "Evolution = tyranny," part of "Formulae for the 21st Century," an Edge/Serpentine Collaboration (2007). Available at http://www.edge.org/3rd_culture/serpentine07/serpentine07_index.html.

7. Daniel G. Gibson et al., "Creation of a bacterial cell controlled by a chemically synthesized genome."

8. Chris Jones, "How to make life," *Esquire* (November 20, 2007). Available at http://www.esquire.com/features/best-brightest-2007/synthbio1207.

9. Jennifer F. Hughes and Steven Rozen, "Genomics and genetics of human and primate Y chromosomes," *Annual Review of Genomics and Human Genetics* 13 (April 5, 2012): 83–108.

10. Steven Rozen, Janet D. Marszale, Kathryn Irenze, Helen Skaletsky, Laura G. Brown, Robert D. Oates, et al., "AZFc deletions and spermatogenic failure: A population-based survey of 20,000 Y chromosomes," *American Journal of Human Genetics* 91 (November 2, 2012): 890–896.

11. François Jacob, "Evolution and tinkering," *Science* 196 (June 10, 1977): 1161–1166.

12. Heinz Neumann, Kaihang Wang, Lloyd Davis, Maria Garcia-Alai, and Jason W. Chin, "Encoding multiple unnatural amino acids via evolution of a quadruplet-decoding ribosome," *Nature* 464 (March 18, 2010): 441–444.

13. Luis Campos, "That was the synthetic biology that was," in *Synthetic Biology: The Technoscience and Its Societal Consequences*, ed. Markus Schmidt, Alexander Kelle, Agomoni Ganguli-Mitra, and Huib de Vriend (Heidelberg: Springer 2009): 5–21.

14. Paul Marks, "Evolutionary algorithms now surpass human designers," *New Scientist* 2614 (July 28, 2007): 26–27.

Chapter 17

1. Bart G. Knols, "On human odour, malaria mosquitoes, and Limburger cheese," *Lancet* 348, no. 9 (1996): 1322.

2. Rachel S. Herz and J. Von Clef, "The influence of verbal labeling on the perception of odors: Evidence for olfactory illusions?" *Perception* 30 (2001): 381–391.

3. Sara Everts, "How advertisers convinced Americans they smelled bad," *Smithsonian*. Available at http://www.smithsonianmag.com/history-archaeology/How-Advertisers-Convinced-Americans-They-Smelled-Bad-164779646.html.

4. Steven Shapin, "Cheese and late modernity," *London Review of Books* (November 20, 2003): 11–15.

5. Heather Paxson, "Post-Pasteurian cultures: The microbiopolitics of raw milk cheese in the United States," *Cultural Anthropology* 23, no. 1 (February 2008): 15–47.

6. The appreciation for the human ecosystem has recently reached a global audience through mainstream press articles like "Modern medicine: Microbes maketh man," *Economist*. Available at http://www.economist.com/node/21560559.

7. Heinrich Anton de Bary, *Die Erscheinung der Symbiose* (Strasbourg, Germany: Karl J. Trübner 1879).

8. Katie Brenner, Lingchong You, and Frances H. Arnold, "Engineering microbial consortia: A new frontier in synthetic biology," *Trends in Biotechnology* 26, no. 9 (September 2008): 483–489.

9. Paxson, "Post-Pasteurian Cultures," 25.

10. Rachel Dutton, "Cheese as a model for the study of microbial ecosystems." Available at http://sysbio.harvard.edu/csb/research/dutton.html.

11. Charles P. Snow, *The Two Cultures and the Scientific Revolution* (Cambridge, England: Cambridge University Press, 1960).

Chapter 18

1. Danielle Pergament, "Nibbling through Spain's cheese country," *New York Times* (November 21, 2008).

2. Mary Douglas, *Purity and Danger: An Analysis of Concept of Pollution and Taboo* (London: Routledge, 2005): 152.

3. Douglas, *Purity and Danger*, 44.

4. Douglas, *Purity and Danger*, 150.

Epilogue

1. Deborah Johnson and Jamison Wetmore, "STS and ethics: Implications for engineering ethics," in *The Handbook of Science and Technology Studies, Third Edition*, ed. Edward J. Hackett, Olga Amsterdamska, Michael Lynch, and Judy Wajcman (Cambridge, Mass.: MIT Press, 2007), 570.

2. Bruno Latour, "A cautious Prometheus? A few steps toward a philosophy of design (with special attention to Peter Sloterdijk)," keynote lecture for the Networks of Design meeting of the Design History Society, Falmouth, Cornwall, September 3, 2008.

3. Johan Schot and Arie Rip, "The past and future of constructive technology assessment," *Technological Forecasting and Social Change,* no. 54 (February–March 1997): 251–268.

4. David Guston, "Innovation policy: Not just a jumbo shrimp," *Nature* 454 (August 2008): 940–941.

5. Matthew Kearnes and Philip Macnaghten, "Introduction: (Re)Imagining nanotechnology," *Science as Culture* 15, no. 4 (December 2006): 279–290 (p. 289).

6. See the Critical Art Ensemble. Available at http://www.critical-art.net/.

7. See http://diybio.org.

8. See the Critical Art Ensemble. Available at http://www.critical-art.net/.

9. Andrew Barry, Georgina Born, and Gisa Weszkalnys, "Logics of interdisciplinarity," *Economy and Society* 37, no. 1 (February 2008): 20–49.

About the Authors

Synthetic Aesthetics Team

Alexandra Daisy Ginsberg is a designer, artist and writer exploring the implications of emerging technologies, and seeking new roles for design. As Design Fellow on the Synthetic Aesthetics project, she has co-curated this international program, developing novel modes of collaboration and critical discourse between art, design and synthetic biology. Other projects experimenting with design in a "biotechnology revolution" include *The Synthetic Kingdom*, a proposal for a new branch of the tree of life; *E. chromi*, a collaboration with Cambridge University's winning team at the 2009 International Genetically Engineered Machine (iGEM) competition; teaching the ArtScienceBangalore and Cambridge University iGEM teams in 2010; and "Synthesis," the design of a pilot six-day laboratory workshop teaching artists, designers and scientists synthetic biology together, produced with The Arts Catalyst, UCL, SymbioticA and Synthetic Aesthetics, funded by the Wellcome Trust. Daisy studied architecture at the University of Cambridge, design at Harvard University, and Design Interactions at the Royal College of Art. Her work has been exhibited at MoMA, the Art Institute of Chicago, The Wellcome Trust, London's Design Museum, the Israel Museum, the National Museum of China, and is in Trento's Museo Delle Scienze's permanent collection. Daisy publishes, teaches, and lectures internationally: talks include TEDGlobal and PopTech. In 2011, *E. chromi* was nominated for the Brit Insurance "Designs of The Year" and the Index Award, and she won the World Technology Award (Design). In 2012, Daisy received the first London Design Medal for Emerging Talent.

Jane Calvert is a reader in Science, Technology and Innovation Studies in the ESRC Innogen Centre at the University of Edinburgh. She has a background in human sciences (Sussex), philosophy of science (London School of Economics), and science policy (Sussex). Before going to university, she trained at London Contemporary Dance School. Jane is particularly interested in the social dimensions of synthetic biology, including interdisciplinary interactions in the field, intellectual property and open source, and the roles of nature, design, and aesthetics in synthetic biology. She is the U.K. principal investigator on the Synthetic Aesthetics project and principal investigator on an ESRC seminar series on synthetic biology and the social sciences. She was a member of the Royal Academy of Engineering's Working Party on Synthetic Biology and the U.K. Synthetic Biology Roadmap Coordination Group. She is currently a member of the Nuffield Council of Bioethics Working Party on Emerging Biotechnologies and the BBSRC Bioscience and Society Panel.

Pablo Schyfter is a lecturer in Science, Technology and Innovation Studies at the University of Edinburgh. His current research focuses on synthetic biology and makes use of sociological and philosophical perspectives on biology, technology, and engineering. Broadly speaking, this research looks to understand the ramifications of bringing to bear engineering on the domain of living nature. A former theatre student, Pablo relished the chance to reengage with art and artists as the Synthetic Aesthetics project's postdoctoral research fellow. In addition to his ongoing work on bioscience and biotechnology, Pablo has carried out research and writing in gender studies and Latin American studies.

Alistair Elfick is co-director of SynthSys—Synthetic & Systems Biology at the University of Edinburgh. Having gained degrees in mechanical engineering and biomedical engineering at the University of Durham, U.K., Alistair won both a Fulbright Commission Distinguished Scholar's Award and a Royal Academy of Engineering Global Research Award, which enabled him to experience biomedical engineering research at the University of California, Berkeley. Having enjoyed too much Californian sunshine, Alistair returned to his hometown of Edinburgh in 2004 to take an EPSRC Advanced Research Fellowship at its eponymous University. The development of novel instrumentation for the measurement of biological samples is the core research undertaken in Alistair's group, work that has served to embed him in the synthetic biology community. Alistair is co-supervisor of Edinburgh's iGEM team and co-authored the Royal Academy of Engineering's inquiry on synthetic biology '09 and has contributed to evidence gathering to inform policy of Scottish and U.K. governments, funding councils, and regulatory bodies.

Drew Endy teaches in the new bioengineering major at Stanford and previously helped start the biological engineering major at MIT. His Stanford research team develops genetically encoded computers and redesigns virus genomes. He co-founded the BioBricks Foundation as a public-benefit charity supporting free-to-use standards and technology that enable the engineering of biology. He co-organized what has become the iGEM competition and the BIOFAB International Open Facility Advancing Biotechnology (BIOFAB). He serves on the U.S. Committee on Science Technology and Law and is a new member of the National Science Advisory Board for Biosecurity. He chaired the 2003 Synthetic Biology study as a member of Defense Advanced Research Projects Agency ISAT Study Group, served as an ad hoc member of the National Institutes of Health (NIH) Recombinant DNA Advisor Committee, and co-authored the 2007 *Synthetic Genomics: Options for Governance* report with colleagues from the Center for Strategic & International Studies and the J. Craig Venter Institute. *Esquire* named Endy one of the 75 most influential people of the twenty-first century. He lives in Menlo Park, California.

Contributors

Christina Agapakis is a synthetic biologist whose research explores the role of design, ecology, and evolution in biological engineering. Her scientific work spans many scales, from proteins to plants to microbial communities. Through collaboration with biologists, engineers, artists, and designers, she pushes the boundaries of synthetic biology from the strict view of cells as computational devices to a more fluid understanding of the function and evolution of living things. In addition to her research and teaching in biological engineering and design, she explores the social and ethical dimensions of biological engineering through writing and film. Having recently completed her Ph.D. at Harvard University, she is currently a postdoctoral researcher in the Department of Molecular, Cell, and Developmental Biology at the University of California, Los Angeles, and a 2012 L'Oreal USA For Women in Science fellow.

David Benjamin is principal at architecture firm The Living and director of the Living Architecture Lab at Columbia University Graduate School of Architecture, Planning, and Preservation. The practice and the lab explore new technologies and create prototypes of the architecture of the future. Recent projects include Living City (a platform for buildings to talk to one another), Amphibious Architecture (a cloud of light above the East River that changes color according to conditions underwater), Living Light (a pavilion in Seoul that displays air quality and collective interest in the environment), and Proof (a series of design studios at Columbia that use evolutionary computation to discover novel, high-performing designs). Before receiving a master of architecture degree from Columbia, Benjamin graduated from Harvard with a B.A. in social studies. He is currently working, in collaboration with Fernan Federici and Tim Rudge (Cambridge University) on the establishment of new educational platforms that combine architecture and synthetic biology.

Will Carey is a designer at IDEO (California) who focuses on combining interaction and industrial design to create provocative and inspiring experiences. He is particularly attentive to the relationship between people and technology and the role of new innovations in shaping our behavior both physically and digitally. He sees an important role for design in provoking thought and discussion, and this curiosity drives him to explore new scenarios in which it is possible to question the social and ethical implications of emerging technology. Will has worked internationally in London, Japan, Milan, and San Francisco. He has exhibited in London, Milan, New York, and Tokyo, and won internationally recognized design awards, including Design Week's Future of Design, Wallpaper's Best Debut Collection, and, most recently, Blueprint's Best Show, during 100% Design London. Will has a B.A. in product design from Central St Martin's, London, and an M.A. in Design Interactions from the Royal College of Art.

Oron Catts is currently setting up a biological art lab called BiofiliA—Base for Biological art and design—at the School of Art, Design, and Architecture, Aalto University, Helsinki, where he is a visiting professor. He is also the founder and director of SymbioticA, The Centre of Excellence in Biological Arts, School of Anatomy, Physiology, and Human Biology, The University of Western Australia, and a visiting professor of Design Interactions, Royal College of Art, London. Oron is an artist, researcher, and curator whose pioneering work with the Tissue Culture & Art Project, which he established in 1996, is considered a leading biological art project. In 2000, he founded SymbioticA, which, under Oron's leadership, won the Prix Ars Electronica Golden Nica in Hybrid Art (2007) and became a Centre for Excellence in 2008. In 2009, Oron was recognized by Thames & Hudson's *60 Innovators Shaping our Creative Future* book in the category "Beyond Design" and by *Icon Magazine* (U.K.) as one of the top 20 designers, "making the future and transforming the way we work." Catts's interest is life, more specifically the shifting relations and perceptions of life in the light of new knowledge and its application. Often working in collaboration with other artists (mainly Ionat Zurr) and scientists, Catts has developed a body of work that speaks volumes about the need for new cultural articulation of evolving concepts of life. Oron was a research fellow at Harvard Medical School and a visiting scholar at the Department of Art and Art History, Stanford University.

Chris Chafe is a composer, improvisor, cellist, and music researcher with an interest in computers and interactive performance. He has been a long-term denizen of the Stanford University Center for Computer Research in Music and Acoustics where he is the center's director and teaches computer music courses. Three year-long periods have been spent at IRCAM, Paris, and The Banff Centre making music and developing methods for computer sound synthesis. The SoundWIRE project launched in 2000 involves real-time Internet concertizing with collaborators the world over. New tools for playing music together and research into latency factors are current goals. He is the Duca Family Professor of Humanities and Sciences at Stanford University. An active performer either on the Net or physically present, his music has been performed worldwide. CDs of works are available from Centaur Records. Gallery and museum music installations are continuing into their second decade with biological, medical, and environmental "musifications" featured as the result of collaborations with artists, scientists, and M.D.s.

Fernan Federici is a postdoctoral researcher at the University of Cambridge working in the area of synthetic biology. He started his career studying two years of engineering at the Universidad Nacional de Cuyo (Argentina) and then moved to Chile to obtain a bachelor's degree in molecular biology (Universidad Catolica, Santiago). He worked for a year at Alvarez-Buylla's lab at Universidad Nacional Autónoma de México (UNAM) and moved to the United Kingdom to obtain a Ph.D. in biological sciences at the University of Cambridge. Fernan is currently working on a project entitled "The Programmable Rhizosphere." This project, co-ordinated by Dr. Haseloff (Cambridge) and Dr. Wipat (Newcastle), seeks to design artificial plant–microbe communication and self-organization.

Hideo Iwasaki works in the fields of both biological science and contemporary art. He obtained his Ph.D. in biology from Nagoya University (1999) and is currently working at Waseda University (since 2005) and is a PRESTO researcher of the Japan Science and Technology Agency (since 2007). As a biologist, he has studied spatiotemporal pattern formation dynamics in cyanobacteria, including molecular genetics of biological clocks, reconstitution of in vitro circadian biochemical oscillations, quantitative analysis of spatial patterning with cell differentiation, and population dynamics of colony pattern formations. As an artist, he has produced contemporary abstract paper cut art to be exhibited as three-dimensional installations and worked on lab biomedia art, especially using cyanobacteria. At his lab, both fine/media artists and scientists share the benches for biology and art simultaneously. His artworks have been exhibited at Havana Biennial, SICF, Holland Paper Biennial, and Artist-in-Residence in Linz. He is co-founder of the Japanese Society for Cell Synthesis Research and the head of the Socio-Cultural Unit.

Mariana Leguia was born and raised in Peru. Throughout her life, she has explored a range of interests including in art, science, and literature. She began her academic training in the Department of Visual Arts at the Catholic University in Lima but soon after was awarded a Fulbright Scholarship and traveled to the United States to study at Lawrence University, where she obtained a B.A. in biology and Spanish literature. Since then, she has been part of research teams at the University of Chicago, Northwestern, Duke, and Brown and has worked on a variety of topics including cancer biology, the biochemistry of heat-shock proteins, and egg activation at fertilization. After obtaining her Ph.D. in molecular biology, cell biology, and biochemistry from Brown University, she moved to the University of California, Berkeley, to do a postdoc in synthetic biology. First she worked on the construction of a minimal synthetic cell derived from mitochondria. Later, she developed foundational technologies for automated, high-throughput assembly of standard biological parts. After 20 years in the United States, she has returned to Peru as the director of molecular biology for the Virology Department of the U.S. Naval Medical Research Unit No. 6, where she is applying the latest technologies to study tropical infectious diseases of public health importance.

Wendell Lim is a professor in the Department of Cellular and Molecular Pharmacology at the University of California, San Francisco (UCSF), and an investigator of the Howard Hughes Medical Institute. Lim received his A.B. in chemistry at Harvard University, his Ph.D. in biochemistry and biophysics at the Massachusetts Institute of Technology, and completed postdoctoral research in molecular biophysics at Yale University. He is the recipient of awards from the Packard Foundation, Searle Foundation, Burroughs Wellcome Fund, and the Protein Society. He is the director of the UCSF/NIH Center for Systems and Synthetic Biology, director of the Cell Propulsion Lab, an NIH Roadmap Nanomedicine Development Center, and deputy director of the National Science Foundation (NSF) Synthetic Biology Engineering Research Center. Lim's research focuses on cell signaling—understanding the molecular circuits that allow cells to communicate, detect signals, make decisions, and execute complex behaviors. He is a pioneer in the emerging field of synthetic biology, which attempts to utilize our understanding of biological mechanisms to engineer cells and biological systems with useful applications in diverse areas ranging from medicine to agriculture to energy. He is a leading expert on how to rewire cells to control and modulate what types of decisions they make. He is interested in how the design process can be creatively applied in science.

Sheref Mansy builds artificial chemical systems that mimic features of cellular life. He studied at Ohio State University and worked on origins of life problems as a postdoctoral fellow at Harvard Medical School and Massachusetts General Hospital. Sheref was awarded a career development award from the Giovanni Armenise–Harvard foundation, which he used to establish a synthetic biology laboratory in Trento, Italy. He is an assistant professor of biochemistry at the University of Trento and a 2012 TEDGlobal fellow.

Sascha Pohflepp is an artist, designer, and writer, born in 1978 in Cologne, Germany. He holds a degree in media art from the Universität der Künste Berlin and an M.A. in Design Interactions from the Royal College of Art in London. He is interested in past and future technologies, notions of art, business, and idealism, what they mean to us, and how they inform which worlds come true and which worlds are discarded. He aims to create social objects in which we can see ourselves differently. For the past five years, he has been contributing to Webby-winning art and technology blog We-Make-Money-Not-Art.com. Most recently, his work has been shown at the Wellcome Trust London, the V&A Museum, and he received a special mention at the 2010 VIDA award for art and artificial life and was nominated for FutureEverything 2010. In 2010, he spent the summer as a research resident at the Art Center College of Design in Los Angeles.

Adam Reineck is a designer with expertise in product design, service design, brand strategy and social innovation. Adam worked for the international consultancy IDEO for 8 years as a principal designer and project leader, tackling challenges as diverse as defining the elite athlete experience for Nike, and building a strategy to help Americans become more energy efficient for the Department of Energy. He was also involved in many of IDEO's first major projects in Asia, and has worked extensively with Samsung, LG, Hyundai, and Hakuhodo in China, Korea, and Japan. In 2012 Adam joined IDEO.org (a non-profit design consultancy) as a Fellow during their foundation year. During this year he worked in Ghana to design new sanitation services in collaboration with Unilever and Water and Sanitation for the Urban Poor, led teams in Mexico developing new low-income financial services with CGAP (a division of the World Bank), and designed ultra-low-cost medical devices currently being tested in Uganda and India. In 2013 Adam founded New Factory, a design studio based in San Francisco. Adam's work has been published in *Fast Company*, *BBC*, *Wired*, *Cell Magazine*, and the *New York Times*, and he has received multiple awards from entities including IDEA, *ID Magazine*, Red Dot, and iF.

Sissel Tolaas has a background in mathematics, chemical science, linguistics and languages, and visual art and has dedicated herself to nose/smell in all levels of life for more than 20 years. She has an archive of 6,730 smells from reality, plus a lab archive of 2,500 molecules. Sissel's knowledge and expertise is in simulation—simulation through synthetic molecules—the air and smells that surround us all the time, from body sweat or smells from hardcore neighborhoods. Her aim is to ask questions, train tolerance, and train awareness. To do this, Sissel works internationally, interdisciplinarily, and collaboratively. Her work is about making systems of smells as a basis for communication, used for the purposes of navigation, education, design, architecture, health care, and environment. Sissel has exhibited at SFMOMA, San Francisco; MoMA, New York; the Guggenheim, Venice and Berlin; Museum of Modern Art, Berlin; National Art Museum of China, Beijing; and at biennales in Berlin, Venice, Tirana, Gwangju, and Liverpool. Recent awards include the Rouse Foundation Award 2009, Harvard Graduate School of Design, and an ArsElectronica Award 2010.

Reid Williams is a biophysics Ph.D. candidate at the University of California, San Francisco, in the laboratory of Wendell Lim. He received B.S. and master of engineering degrees from MIT, studying electrical engineering and computer science. His research at UCSF focuses on using synthetic biology to understand the design principles of biological circuits, which underlie fundamental cellular behaviors like cell replication, memory formation, and tissue development. This includes building tools to engineer one circuit mechanism, based on protein phosphorylation, and studying cellular behaviors that arise from combinations of this and other circuit types.

Ionat Zurr is an award-winning artist and researcher who formed, together with Oron Catts, the Tissue Culture & Art Project. She has been an artist in residence in the School of Anatomy, Physiology, and Human Biology at the University of Western Australia since 1996 and was central to the establishment of SymbioticA in 2000. Zurr, who received her Ph.D. titled "Growing Semi-Living Art" from the Faculty of Architecture, Landscape, and Visual Arts at the University of Western Australia, is a core researcher and academic coordinator at SymbioticA. She is considered a pioneer in the field of biological arts, and her research been published widely, exhibited internationally, and her artwork has been collected by MoMA, New York. Ionat is a recipient of the Discovery Australian Research Council Award (2012); she has been a fellow in the InStem Institute, NCBS, Bangalore (2010), and a visiting scholar at The Experimental Art Center, Stanford University (2007), and The Tissue Engineering & Organ Fabrication Laboratory, Massachusetts General Hospital, Harvard Medical School (2000–2001). She has exhibited in places such as MoMA New York, Mori Museum Tokyo, Ars Electronica Linz, GOMA Brisbane, and more. She is currently a visiting professor at Future Art Base, School of Art, Design, and Architecture, Aalto University, Helsinki, Finland.

Source Notes

Every effort has been made to contact the original copyright holders of the images in this book. Any rights holders not credited should contact the publisher.

Introduction

Figure I.1

E. coloroid. Photograph courtesy of Chris Voigt (MIT) and University of Texas, Austin and University of California, San Francisco 2004 iGEM team. Photograph by Aaron A. Chevalier, 2004.

Figure I.2

Bacillus subtilis. Photograph by Fernan Federici, 2012.

Figure I.3

E. glowli. By University of Cambridge iGEM team 2010. Photograph by Fernan Federici, 2010.

Figure I.4

Intolerable Beauty: Portraits of American Mass Consumption, Crushed Cars #2, Tacoma 2004. Photograph by Chris Jordan, 2004.

Figure I.5

Algae Bloom. Copyright © Reuters, 2011.

Figure I.6

Christina Agapakis's Lab Bench, Harvard. Photograph by Alexandra Daisy Ginsberg, 2010.

Figure I.7

Sissel Tolaas's Studio, Berlin. Photograph by Alexandra Daisy Ginsberg, 2010.

Chapter 1

Frontispiece

Bottles of ATCG. Photograph by Theo Cook. Art direction by Kellenberger-White, 2012. Bottles provided by Link Technologies Ltd., Scotland.

Figure 1.1

Silicon Wafer. Courtesy of the Intel Corporation.

Figure 1.2

Three Stages of Pinecone Growth. Photograph by Cavan Images, 2007. Copyright © Cavan Images/The Image Bank/Getty Images.

Figure 1.3

Tree of Life, Materials and Made-Life. Illustration by Kellenberger-White, 2012.

Figure 1.4

BioBrick Assembly. Illustration by Kellenberger-White, 2012.

Figure 1.5

Horsepower and Equivalents. Illustration by Kellenberger-White, 2012.

Figure 1.6

Synthetic Biology Abstraction Hierarchy. Illustration by Kellenberger-White, 2012.

Figure 1.7

Design–Build–Test Cycle. Illustration by Kellenberger-White, 2012.

Figure 1.8

Oil Refinery at Night. Photograph by Wolfgang Schlegl, 2007.

Figure 1.9

Geneforge Coliseum High-Throughput Synthesizer. Photograph courtesy of Geneforge Inc.

Figure 1.10

The International Genetically Engineered Machine competition, 2010. Photograph courtesy of iGEM & Justin Knight.

Chapter 2

Frontispiece

Evolution of the Engineering Mindset Toward Life. Photograph by Theo Cook. Art direction by Kellenberger-White, 2012.

Figure 2.1

Jacques Loeb's Artificial Life. Headlines and illustration from *Chicago Daily Tribune*, December 28, 1900.

Figure 2.2

Miller–Urey Experiment. Photograph courtesy of Mandeville Special Collections Library, University of California, San Diego.

Figure 2.3

Microvenus. Illustration by Kellenberger-White, 2012, based with permission on "Microvenus" by Joe Davis, 1986.

Figure 2.4

Victimless Leather. Project and photograph by the Tissue Culture & Art Project (Oron Catts and Ionat Zurr), 2004.

Figure 2.5

Tissue Engineered Steak No. 1. Project and photograph by The Tissue Culture & Art Project (Oron Catts and Ionat Zurr), 2000.

Chapter 3

Frontispiece

Greening Biotechnology. Photograph by Theo Cook. Art direction by Kellenberger-White, 2012.

Figure 3.1

Amyris's Engineered Yeast Producing Farnesene as Seen Under a Microscope. Photograph courtesy of Amyris Inc., 2010.

Figure 3.2

Shagreen Skin. Photograph by Melissa Tedone. Courtesy of Iowa State University Library, 2010.

Figure 3.3

Teosinte to Maize. Courtesy of Robert S. Peabody Museum of Archaeology, Phillips Academy, Andover, Massachusetts.

Figure 3.4

PostNatural Organisms of the Month. Courtesy of the Center for PostNatural History, 2012.

Figure 3.5

The Synthetic Kingdom. Project and illustration by Alexandra Daisy Ginsberg, 2009.

Figure 3.6

The Synthetic Kingdom: Iterations. Project and illustrations by Alexandra Daisy Ginsberg, 2010.

Figure 3.7

Le Corbusier's Hands. Still from Pierre Chenal' s film *Architecture d'Aujourd'hui*, 1930, reproduced with permission of the Fondation Le Corbusier, Paris, and Artists' Rights Society (ARS). New York/ADACP.

Figure 3.8

First Bacterial Genome Transplantation. Photograph reproduced with permission of J. Craig Venter Institute, 2010.

Figure 3.9

Diurnisme. By Décosterd & Rahm Associés, 2007. Photograph by Adam Rzepka/Pompidou Centre, 2007.

Figure 3.10

a/b. Illustration by Kellenberger-White, redrawn with permission from illustration by Dunne & Raby, 2009.

Figure 3.11

Acoustic Botany: Lab Bench Test Rig. Image by David Benque, 2010.

Figure 3.12

Microbial Kitchen. By Clive van Heerden and Jack Mama, Philips Design Probes, Philips, 2011.

Chapter 4

Frontispiece

Nature Is Designed. Photograph by Theo Cook. Art direction by Kellenberger-White, 2012.

Figure 4.1

Bacteriophage øX174. Diagram by Drew Endy, 2012.

Figure 4.2

Bacteriophage øX174 Genome Map. Illustration by Kellenberger-White, 2012.

Figure 4.3

Alien Messages in øX174 Gene B. "Is Bacteriophage øX174 DNA . . . ," paper by Hiromitsu Yokoo and Tairo Oshima, *Icarus* 38, 148–153 (1979); "Scientists Examine Tiny Viruses . . . ," article by Walter Sullivan, *New York Times,* May 7, 1979.

Figure 4.4

Fly Eye. Photograph by Kelby Douglas, 2012.

Figure 4.5

DNA's Structure. Illustration by Kellenberger-White, 2012.

Figure 4.6

The Genetic Code. Illustration by Kellenberger-White, 2012.

Figure 4.7

Venus Flower Basket Sponge. Photograph by Ryan Moody, 2008.

Chapter 5

Frontispiece

Delphinium Spectrum. Photograph by Theo Cook. Art direction by Kellenberger-White, 2012.

Figure 5.1

Motorbike Sound Engineering. Photograph courtesy of AVL List GmbH.

Figure 5.2

Mangrove Roots. Photograph by Kate Dagget, 2008.

Figure 5.3

Cat Fancy Club. A Somewhere project by Nina Pope and Karen Guthrie. Photograph by Marc Henrie. Image compositing by Theo Cook, 2010.

Chapter 6

Frontispiece

E. chromi. Photograph by Theo Cook. Art direction by Kellenberger-White, 2012. *E. chromi: The Scatalog* by Alexandra Daisy Ginsberg and James King with the University of Cambridge 2009 iGEM Team, 2009–2011.

Figure 6.1

Fanged Pitcher Plant. Photograph by David Burwell, 2009.

Figure 6.2

Growth Assembly: Herbicide Gourd. Alexandra Daisy Ginsberg and Sascha Pohflepp. Illustration by Sion Ap Tomas, 2009.

Figure 6.3

Growth Assembly: Nozzle Fruit. Alexandra Daisy Ginsberg and Sascha Pohflepp. Illustration by Sion Ap Tomos, 2009.

Figure 6.4

Growth Assembly: Assembled Herbicide Sprayer. Alexandra Daisy Ginsberg and Sascha Pohflepp. Illustration by Sion Ap Tomos, 2009.

Figure 6.5

Mating Aedes aegypti. Photograph by Said Bounodi, Oxitec Ltd, and Paul Reiter (Pasteur Institute, France), 2009.

Figure 6.6

Fertilized Cow Eggs. From *The Journal of Controlled Release*, Volume 150, Issue 1, February 28, 2011, pages 23–29. Reproduced with the permission of Elsevier. Photograph by Fussenegger Lab.

Figure 6.7

BioCouture: Bio Ruff Jacket. Suzanne Lee. Commissioned by Science Museum London. Photograph by Santiago Arribas, 2010.

Figure 6.8

Mycotecture Brick. Project and photograph by Phil Ross, 2009.

Figure 6.9

BioConcrete Testing. Photograph courtesy of Henk Jonkers, 2008.

Figure 6.10

Spirulina Culture Ponds at Cyanotech Corporation in Kona, Hawaii. Photograph reproduced by permission of Cyanotech Corporation, 2011.

Figure 6.11

Biomanufactured Brick. Verge Labs. Photograph by Ginger Krieg Dosier, 2010.

Figure 6.12

E. chromi: The Scatalog. Alexandra Daisy Ginsberg and James King with the University of Cambridge 2009 iGEM Team, 2009–2011. Photograph by Åsa Johannesson.

Figure 6.13

The Synthetic Kingdom: Keratin Cup. Alexandra Daisy Ginsberg, 2009. Photograph by Carole Suety.

Figure 6.14

The Synthetic Kingdom: Carbon Monoxide Detecting Lung Tumour. Alexandra Daisy Ginsberg, 2009. Photograph by Carole Suety.

Figure 6.15

Common Flowers/Flower Commons. BCL/Shiho Fukuhara and Georg Tremmel. Photograph by Georg Tremmel, installation at the Wallraf-Richartz Museum, Cologne, 2008.

Figure 6.16

Searching for the Ubiquitous Genetically Engineered Machine. Project and photograph by ArtScienceBangalore, 2011.

Figure 6.17

IDEO Bacterial Photography. Will Carey, Adam Reineck, and Reid Williams. Photograph by Reid Williams, 2011.

Chapter 7

Frontispiece

Artichoke Structural Logic. Photograph by Theo Cook. Art direction by Kellenberger-White, 2012.

Figure 7.1

Artichoke Xylem Tissue. Photograph by Fernan Federici, 2011.

Figure 7.2

Artichoke Xylem Cell. Photograph by Fernan Federici, 2011.

Figure 7.3

DEX-induced Xylem Formation. Photograph by Fernan Federici, 2011.

Figure 7.4

Trapped Protoplast. Photograph by Fernan Federici, 2011.

Figure 7.5

Mathematical Modeling and Design Application. Illustration by David Benjamin, 2011.

Figure 7.6

Cell and Logic Prototypes. Photograph by David Benjamin, 2011.

Figure 7.7

E. coli Biofilm Patterns. Photograph by Fernan Federici, 2011.

Figure 7.8

Composite Sheets: Computational Evolution. Illustration by David Benjamin, 2011.

Chapter 8

Figure 8.1

Peter Cook and Colin Fournier's Kunsthaus Graz. Courtesy of Kunsthaus Graz; photograph by Elvira Klamminger.

Figure 8.2

Generated Turing Patterns. Illustration by Shigeru Kondo, from "Reaction-diffusion model as a framework for understanding biological pattern formation," Shigeru Kondo and Takashi Miura, *Science* 329, 1616 (2010). Reprinted with permission from the American Academy of Arts and Sciences.

Figure 8.3

Slime Mold Computation. Photograph by Dr. Seiji Tagaki, from "Rules for biologically inspired adaptive network design," Atsushi Tero et al., *Science* 327, 439 (2010). Reprinted with permission from the American Academy of Arts and Sciences.

Chapter 9

Frontispiece

Night Science. Photograph by Theo Cook. Art direction by Kellenberger-White, 2012.

Figure 9.1

Yeast Mutation. Photograph by Will Carey, 2011.

Figure 9.2

IDEO Design Process. Illustrated by Kellenberger-White, 2012. Adapted with permission from illustration by IDEO.

Figure 9.3

Living among Living Things Concept Sketch. Illustration by Will Carey and Adam Reineck, 2011.

Figure 9.4

Personal Microbial Culture 1. Will Carey and Adam Reineck. Photograph by Nicolas Zurcher, 2011.

Figure 9.5

Personal Microbial Culture 2. Will Carey and Adam Reineck. Photograph by Nicolas Zurcher, render by IDEO, 2011.

Figure 9.6

Packaging Creates Contents. Will Carey and Adam Reineck. Photograph by Nicolas Zurcher, render by IDEO, 2011.

Figure 9.7

Science Process. Illustration by Kellenberger-White. Adapted with permission from a diagram by Wendell Lim, 2011.

Chapter 10

Figure 10.1

Tower Bridge Alternatives. Nineteenth-century print of Horace Jones' bridge design. Courtesy of the London Metropolitan Archives.

Figure 10.2

Glass Tower Bridge. Illustration by W.F.C. Holden, 1943. Courtesy of the London Metropolitan Archives.

Figure 10.3

Professional Hierarchy Rethought. Illustration by Kellenberger-White, 2012.

Chapter 11

Frontispiece

Fossilized Stromatolite. Photograph by Theo Cook. Art direction by Kellenberger-White, 2012.

Figure 11.1

Oscillatoria sp. Photograph by Hideo Iwasaki, 2012.

Figure 11.2

Geitlerinema sp. Photograph by Hideo Iwasaki, 2012.

Figure 11.3

Lake Clifton Thrombolites. Photograph by Perdita Phillips, 2009.

Figure 11.4

Australian Mining. Photograph by MK Photography, 2007.

Figure 11.5

Fool's Gold. Photograph by Anthony Swinehart.

Figure 11.6

Cyanobacteria Digesting Printed Computer Boards. Hideo Iwasaki. Installed at the *BioAesthetics* exhibition, Tokyo. Photograph by Hideo Iwasaki, 2012.

Chapter 12

Figure 12.1

Geological Timescales. Illustration by the U.S. Geological Survey. Spiral designed by Joseph Graham, William Newman, and John Stacey. Digital preparation by Will Stettner.

Figure 12.2

Growing Thrombolites. Photograph by Oron Catts, 2010.

Figure 12.3

Between Reality and the Impossible. Dunne & Raby. Photograph by Jason Evans, 2010.

Chapter 13

Frontispiece

Sonifying DNA. Photograph by Theo Cook. Art direction by Kellenberger-White, 2012.

Figure 13.1

Orbiting Gear Shift: DNA Sonified. Illustration by Kellenberger-White, 2012.

Figure 13.2

Room Reverberations. Illustration by Kellenberger-White, 2012.

Figure 13.3

Acoustic Impulse Responses. Illustration by Kellenberger-White, 2012.

Figure 13.4

DNA Chromatogram. Illustration by Kellenberger-White, 2012.

Figure 13.5

DNA Rooms. Illustration by Kellenberger-White, 2012.

Chapter 14

Figure 14.1

DNA Code. Illustration by Kellenberger-White, 2012.

Figure 14.2

Pablo's Brain Waves. Printed with permission of Pablo Schyfter, 2002.

Figure 14.3

DNA Sonification. Diagram by Chris Chafe, 2011.

Figure 14.4

Gels. Photograph by Melanie Jackson, Synthesis Workshop, 2011.

Chapter 15

Frontispiece

Steam/Metabolism. Photograph by Theo Cook. Art direction by Kellenberger-White, 2012.

Figure 15.1

An Ox Treadmill. Woodcut from Zonca's *Novo Teatro di machine et edifici,* 1607.

Figure 15.2

Fowler's Steam Engine. Illustration from William & Robert Chambers, *Encyclopaedia—A Dictionary of Universal Knowledge for the People* (Philadelphia: J.B. Lippincott & Co., 1881).

Figure 15.3

Ideal Horse. Woodcut by Erhard Schön, taken from *Unterweisung der Proportion und Stellung der Possen* [Nürnberg, 1542; In getreuer Nachbildung herausgegeben mit einer Einführung von Leo Baer]. Frankfurt a.M.: J. Baer, 1920.

Figure 15.4

Centrifugal Governor. Illustration by William Dwight Whitney, 1911. William Dwight Whitney, *The Century Dictionary: An Encyclopedic Lexicon of the English Language* (New York: The Century Co., 1911).

Figure 15.5

Geometric Human Figureure. Woodcut by Erhard Schön, taken from *Unterweisung der Proportion und Stellung der Possen* [Nürnberg, 1542; In getreuer Nachbildung herausgegeben mit einer Einführung von Leo Baer]. Frankfurt a.M.: J. Baer, 1920.

Figure 15.6

Lipid Vesicles. Photograph by Sheref Mansy, 2011.

Figure 15.7

Dallinger's Incubator. Royal Microscopical Society. Unknown Illustrator. From W.H. Dallinger, "The President's Address," *Journal of the Royal Microscopical Society* (1887): 185–199.

Chapter 16

Figure 16.1

Mule Family Tree. Illustration by Kellenberger-White, 2012.

Figure 16.2

Evolution = Tyranny. Illustration by Drew Endy, 2007. Drawn at "Formulae for the 21st Century," Edge Live in London/The Serpentine Gallery Experiment Marathon, October 2007.

Figure 16.3

Human X and Y Chromosomes. Scanning electron microscopy by Biophoto Associates/Science Photo Library.

Chapter 17

Frontispiece

Body/Cheese. Photograph by Theo Cook. Art direction by Kellenberger-White, 2012.

Figure 17.1

Cheese Varieties. Photograph by Gemma Lord, 2012.

Figure 17.2

Molecules of Cheese Chemistry. Illustration by Kellenberger-White, 2013.

Figure 17.3

Collecting Bacteria Samples. Photograph by Kellenberger-White, 2012.

Figure 17.4

Human Cheese Making 1: Sissel and Christina. Photograph by Alexandra Daisy Ginsberg, 2010.

Figure 17.5

Human Cheese Making 2: Bottles. Photograph by Alexandra Daisy Ginsberg, 2010.

Figure 17.6

Human Cheese Making 3: Curds. Photograph by Alexandra Daisy Ginsberg, 2010.

Figure 17.7

Human Body Cheese. Photograph by
Kellenberger-White, 2012.

Figure 17.8

Nose-2 Smell Trace. Illustration by
Kellenberger-White, 2012.

Figure 17.9

Microbial Diversity. Photograph by
Alexandra Daisy Ginsberg, 2010.

Chapter 18

Figure 18.1

Cheese Making by Hand. Photograph by
Matias Costa for *The New York Times*/
Redux/eyevine, 2008.

Figure 18.2

Breath Testers. Photograph copyright
© Louie Psihoyos/Science Faction/Corbis.

Figure 18.3

The Human Microbiome. Illustration by
Kellenberger-White, 2012, adapted from
illustration by Darryl Leja, National
Human Genome Research Institute, 2009.

Figure 18.4

Cloaca New & Improved. Wim Delvoye,
2001. Photograph by Studio Wim Delvoye,
The Power Plant, Exhibition, 2004.

a/b: a changing understanding of design (Dunne & Raby), 63, *65*

Abalone snail shell strength, 6

Abiogenesis, 30

Abstraction, 231–233
    as form of representation, 238–239
    hierarchy in synthetic biology, *15*
    promoter as faucet, 14–15
    and representation, 231–242
    and withdrawal, 241–242

Acoustic Botany (Benque), 66–67, *67*

Acoustics
    importance in motorcycle engine design, 89
    simulations, 220, 225, *226*, 227, *228–229*

Aesthetics. *See also* Art; Design; Synthetic
    Aesthetics Residences
    "aesthetic criteria" used by Poincaré, 184
    aesthetic pleasure, designing solely for, 66, 129
    aesthetic sensibilities, 174
    issues raised by new materials and
        experiences, 62–63
    motorcycle design, aesthetic aspects, 88–89
    and values, 96

Agapakis, Christina, *xix*, 296–297, 301. *See also*
    The Inside-Out Body project
    lab bench, *xix*

Age of the Anthropocene, 197

Algae, 122
    bloom triggered by water pollution, *xvi*
    commercial production, Australia, 121–122,
        121, 200

Algorithms
    evolution as algorithm to select optimized
        genetic designs, 16
    genetic, 109, 143

Alien messages, bacteriophage øX174, news
    reports, *77*

Alon, Uri, 179

Amino acids, linked to specific DNA triplets,
    82–83, *83*

Analogies, use in synthetic biology, x, 11, 18, 162,
    234, 239

Analysis, for olfactory science, *287*

Angerer, Daniel (chef), 290

Animals. *See also* Cat Fancy Club; Dogs; Horses;
    Mules
    competitive advantage, 78–79
    domestic, designed, *51*, *52*
    as machines, 248–251, *249*

Anthropocene, Age of the, 197

Antonelli, Paola (Senior Design Curator at
    MoMA), 63–65, 212

Appearance, cheese bacteria, *279*

Apple Inc., aesthetic sensibility, 174

Applications, design, as bioengineer's focus, 40

Arabidopsis thaliana
    DEX-induced xylem formation, *147*
    leaf, *146*

Architecture
    and biology, 156–159
    computational tools, 143–144
    manufacture and construction,
        transformational possibilities, 153
    modernist, 57
    scale, 147–148
    Synthetic Aesthetics project as educational
        model of design studios, 154

Arduino platform, 170

"Armpit" cheeses
    bacteria isolated, *279*
    "Daisy Armpit" cheese, *278*, 285
    odors isolated, *279*
    smell notes, *279*

Arnieva, Oliva Peláez (cheese maker), 283–285

Arnold, Frances (bioengineer), 28

Arsenic, 6
  biosensor development, 133
  detecting in drinking water, 7, 131
Arsenic Biosensor (U of Edinburgh iGEM entry),
  131–133
Art. *See also* Bio art
  and critique, 210–211
  or design, 211–214, 261–262
  and human body fluids, 288, 290
  involvement in early stages of development of
    synthetic biology, 299
  process, 40
  rearrangement of matter, 195–204
  role in disciplinary mix, 210–211
  and synthetic biology, 27–37, 212
  and utility, 70, 211
Artichoke
  structural logic, *140*
  vascular tissue, *145*
  xylem cell, *145*
"Artifact," different meanings, 40
Artificial products, made from a single natural
  source, 50
Artistic research, places of biological and
  biogenic geological significance, 201
Artists
  and engineering mindset, 37
  and scientists working together, 210
ArtScienceBangalore, iGEM projects, 33, 131, *132*
"As if ..." (Dunne & Raby), 67
*ATCG, bottles of (the four nucleotide bases of
  natural DNA)*, 2
Aurora Algae, Karratha (commercial algae site),
  200
The Autotroph, 204

Baby Gaga ice cream, 290
Bacteria
  Bacillus subtilis engineered by Fernan
    Federici, *xii*
  bacteria-based manufacturing, *176*
  Brevibacterium linens and flavor of cheese, 273
  in cheeses, isolation and sequencing
    techniques, *279*
  and disease, connection, 287
  engineered, *xiii*
  engineering to produce commercial products,
    254
  and flavor of cheeses, 272–273
  from different human cheese, *279*
  importance in cheese making, 272–273
  in production of cellulose, 115
  smell notes, *279*
  to produce flavor of cheese, 273
Bacterial patterns
  biofilm, as tool for biofabrication, 151–154
  as manufacturing tools, 144

Bacteriophages, 74–75
  øX174 bacteriophage, *74, 76*
Beadle, George, "one gene, one enzyme"
  hypothesis, 75
Beauty, and utility, 214
Benjamin, David, 108, 134–135, 296, 301. *See
  also* Bio Logic project
Benner, Steven, 82
Bennett, Gaymon, 133
Benque, David, *Genetically Engineered Sound
  Garden*, 66, *67*
Berlin cheese shop, 285
  professional cheesemongers' opinions, *279*
"Better"
  design, 56, 63
  meaning, xvii, 62
  synthetic biology, 298–299
Better Brick process, 122–123, *123*
Between Reality and the Impossible, Future Foragers
  (Dunne & Raby), *213*
Bialek, Bill, 80
BioAesthetics exhibition, Tokyo, cyanobacteria
  dissolving printed computer boards, *202*
Bio-analogs, 18
Bio art, 69–70, 211
BioBricks, 12–13, *12*
  and "blackboxing," 16
  use in iGEM competition, 33, 123–124, 131,
    132
BioBricks Foundation, 130
"Bio CAD," 45
Biocomputing, 159–161, 163
  and logic of biology, 160–161
BioConcrete, 118–119, *118*
BioCouture, 115, 117
  jacket, *116*
Bio-digester Kitchen Island, 68
Biodiversity
  and fuel needs, 122
  nature, constitutional rights, 130
Bioengineering, and Biotechnology Revolution,
  x–xi
Bioerror, 29, 59–60
Bioethics, 66, 97
Biofabrication. *See also* Biomanufacturing
  bacterial biofilm patterns as tool for, 152–154
Biofabricator (iGEM project), 103
Biogenic formations, human-induced, 201
Biogenic Timestamp project, xxii, 195–204, 206,
  208–214
Biological design
  context-aware, 136
  critically engaged, 298
  and evolution, 80, 136, 255–256
  eye of fly, *79, 80*
  model for good practice, 108–109, 122
  new framework, 154

new paradigm, 108
as novel space for critical discourse and
practice, 134
potential, 107, 118–119
Venus flower basket sponge, *84*
Biological engineering, 9, 39
16th C, of the animal body, 251
distinction from other kinds of engineering,
296
Biological things, possibility of living with, 299
Biological time, and geological time, 198, 206
Bio Logic project, xxi, 139–164. *See also* Logic
of biology
Biology. *See also* Nature
and architecture, 156–159
containment, 128
as direct design and manufacturing tool, 144
and engineering, relationship, 282
harnessing complexity, 18
nature of, 296–298
unique properties, 296–297
use to manufacture human artifacts, 6
wilfulness and malleability, 188
Biomanufacturing. *See also* Biofabrication
disruptive, 59, 103
and synthetic biology, 115
Biomass, sustainable materials, 120
"Biomimicry," 156, 162–164
Bio-similars, 18
Biotechnology
as disappointment, 19
industrial, and synthetic biology, 18
products from, 19–20
simplifying and deskilling, 18
unexpected outcomes, 59
Biotechnology Revolution
and bioengineering, x–xi
shaping of the 21st century, ix–x
Bioterror, 59–60
"Black-boxing," 15–16, 202, 256
and BioBrick standard, 16
The Blob (film, 1958), 60–61
Blue-green algae. *See* Cyanobacteria
Blurring boundaries, abstraction hierarchies, *190*
Bodies/the body. *See also* Human body
and agricultural steam engines modeled on
bodies of horses and oxen, *250*
Inside-Out body, 269–293
as malleable entity, 29–31
as places of manufacture, *124, 126–127,* 128
power of living body, 258
as raw material, 34–35, 37
as site for energy conservation, 248
Body/Cheese, 268
Body fluids, human, and art, 288, 290
Body odor, human, and deodorization, 274

Boron, 6
*Bottles of ATCG, the four nucleotide bases of natural
DNA, 2*
"Bottom-up" systems biology, 23, *58,* 254. *See also*
Protocells; "Top-down"
and architectural approach, 57
Boundaries, transgressing biological, 283–293
Boyd, Dana (molecular biologist), 32
Boyle, Patrick (biologist), 122
Breeding, as aesthetic choice, 83, 96
Brevibacterium linens, and flavor of cheese, 273
Brie cheese, *272*
fungi in production, 273
Building materials
Better Brick process, 122–123, *123*
BioConcrete, 118–119, *118*
generated through bacterial manufacturing,
153
"local materials," 153
made from natural products, 6
mushrooms as, 117, *117*
Mycotecture brick, *117*
timber as alternative to steel, 115

Caenorhabditis elegans (worms), 131
Calvert, Jane, xviii
Cambridge University. *See also* Federici, Fernan;
Haseloff, Jim
biosensor development, 133
Jason Chin's lab, 54
iGEM teams (*see E. chromi; E. glowli*)
synthetic biology students' *E. glowli, xiii*
Cannibalism, and food, 292
Carbon dioxide, increased levels on Earth, 197
Carbon Monoxide Sensing Lung Tumour, (*The
Synthetic Kingdom*), *127*
Carey, Will, 135, 136, 297, 299. *See also* Living
among Living Things project
Carrel, Alexis, 30–31
Cat Fancy Club, 95, 96
Catts, Oron, 43, 69–70, 136, 262, 297–299, 301.
See also Biogenic Timestamp project
(with Alexandra Daisy Ginsberg), "The Well-
Oiled Machine" (short story), 114
Cells, as factories, 10
Cellulose
durability, 117
kombucha to make, 115, *116*
Center for Post-Natural History, 52
PostNatural Organisms of the Month, *51*
Chafe, Chris, 136, 296. *See also* Synthetic Sound
from Synthetic Biology project
Change
avoidance in synthetic biology, 263
and evolution, 262
insulating devices from, 16
of organisms over time, and design, 118, 256

Chassis, 22–23, 54, 62, 232
  as analogy, 11, 234, 239
  complexity, 16
Cheese. *See also* Human cheeses
  bacteria on cheese makers' hands, 283–284
  curds of "human cheeses," *277*
  and disease, 281
  as model organism, 282
  production, 271–272
  smells, *279*
  varieties, *272*
Cheeses, human, *278*
Chemical plant, *21*
Chemistry, 254
  of cheese, important molecules, *273*
China National Stadium, 115
Chin, Jason, 54
Christopher, John, *The Death of Grass*, 60
Chromatograms, 225, 227, *229*. *See also* Gas
  chromatography
  and acoustic properties of rooms, *228–229*
Chromosomes
  human Y chromosome, *266*
  modification to operate for one generation
    only, 263–264
Churchill, Winston, on growing meat, 34–35
Circadian timekeeping, 203, 206
"Circuits," 232
Circuits, genetic, 18
  creation, 220–221
Climate change, 47, 122, 204
  effect on thrombolites, 201
Clock genes, 203
Cloud, fear of, 179
Coding region (of gene), 11
Codons. *See Translation start and stop codons*;
  "Triplet codes"
Collaboration
  between scientists and designers. *See*
    Interdisciplinary collaborations
  as interdisciplinary practice, 133
Colors. *See also E. chromi* collaborative project
  bacteria to produce, 123–124
  and *The Scatalog*, 125
  and *The Synthetic Kingdom*, 128
Columbia University Graduate School of
  Architecture, Planning and Preservation, 154
Common Flowers/Flower Commons, 129, *129*
Complexity, 163
"Compromise solution," 109
Computational evolution of composite sheets,
  153
Computational modeling, 17–18
Computational techniques
  architecture and synthetic biology, 157–158, 160
  used in translation of exoskeleton structures
    to architecture, 141–147–8

Computational tools, architecture, 143–144
Computer chips, *4*
Computer music
  and synthetic biology, 230
  techniques, 220
Computer programming language terms, used
  in synthetic biology, 240
Computer scientists' involvement in synthetic
  biology, evolutionary computing, 143
Computing analogy, x
Concert hall reverberations, simulations, 220
Confocal microscopy, 159
Consequences of new technical discoveries,
  Danielli on, 58
Construction materials. *See* Building materials
Consumption, reduction, lack of impact, 115
Context
  and biological design, 136
  as design tool, 119
  and disgust, 290–291
  importance, 290–291, 293
Control, and design, 265
Cooperation
  between human designer and living cell, 151
  as interdisciplinary practice, 133
Copyright, 130
Creativity, 46, 184
  constrained from engineering perspective,
    181–191
  and evolution of design, 184
  fostering, by Google corporation, 189
  iGEM projects, 189–190
  importance, 297
  as possible engine of capitalism, 187–188
  in science, 176–177, 179–180
  and synthetic biology, 188, 297
Crick, Francis, 9–10, 78
Critical design, 63, 66, 136, 212
  and art, 212–213
Critique, 45–46
  and art, 210–211
Crowley, David (design historian), 64
"Crushed Cars #2, Tacoma 2004," *xv*
Cyanobacteria
  biological clock, manipulating, 195
  cell cycle, 199
  as dominant life form, 196
  Oscillatoria cyanobacterium, *196*
  potential usefulness, 206
  relevance across timescales, 206
  spatiotemporal pattern dynamics, 203
  time-dependency, 195, 199
  use for human purposes, 197
CyanoBonsai, 204
Cybernetics, 252–253
  and engineering life, 251–252

Dallinger, Rev. Dr. William, 256
  incubator, *257*
Damage, caused by biotechnologies, xi
Danielli, James F., 58, 61
DARPA (Defense Advanced Research Projects
  Agency) (US), 60
Darwin, Charles. *See also* Evolution
  designing with Darwin, 255–256
  on selection by humans, 53
Data, as representation, 233–234
Davis, Joe (artist), 33
  Microvenus (first nonbiological message
    encoded in DNA), *32*
Day science, and night science, 176–177
The Day of the Triffids (Wyndham), 60
Death. *See* Life and death; Product death
Decomposable materials, 115–118
Décosterd, Jean-Gilles (architect), 63
  Diurnisme (2007), *64*
"Deep-dive" workshop, 171–172, 178
Defense Advanced Research Projects Agency
  (DARPA) (US), 60
Defense Threat Reduction Agency (DTRA)
  (US), 60
Definitions. *See also* Language
  of boundaries, ethical design, 112
  design, 41, 43–45, 88, 211
  NASA definition of life, 262
  sonification, 219
DeLanda, Manuel, 203, 206
Delphiniums, *86*, 96–97
Delphinium Spectrum, 86
Delvoye, Wim, *Cloaca New and Improved*, 288,
  290
de Mora, Kim, 131–132
Dengue fever, 110
Deodorization, 274
de Rijke, Alex (architect), 115
Design. *See also* Biological design; Ethical design
  or art, 211–214, 261–262
  of biology, 82–83
  by people and only by people, 87
  change of function, 44–45
  and change over time, 118
  constraints on, 178
  contemporary mindset, 43
  contrasts with synthetic biology, 65–66
  crit or charette, 42, 122
  definitions, 41, 43–45, 88, 211
  different meanings, 40–41
  and evolution, 78–79, 91, 93, 113, 259–267,
    296
  evolution of, 100–137
  future designers, 45
  involvement in early stages of development of
    synthetic biology, 299
  lack of design in nature, 87

and long-term thinking, 47
as medium to trigger debate and discussion, 63
modernist perspective, 57
in nature, 47, 78–79, 90–91, 93, 259–267
as new perspective, 63
and "night science," 178–179
principles, 108
process, 40, 178
professional designers, function, 46
reinvention possibility, 70
revolution, 63
and synthetic biology, xiv, 40, 43, 45, 93–95,
  97, 107
and technology, 187–188
Design/build/test cycle, 17, *17*
"Design crit," Boyle's workshop for synthetic
  biology design, 122
"Design for debate." *See* Critical design
Designer
  role of, 41, 63
  as social critic, 66
Design fictions, 66–67, *67*
Designing with Darwin, 255–256, 296
Design mindset, xx–xxi, 45
"Deskilling"
  biotechnology, 18
  the design process, 45, 188
Detritus, of today, *xv*
Developing world, synthetic biology in, 24–25
"Devices," 232–233
de Vries, Hugo, 265
Dewey, John, 232
Directed evolution, 16–17, 24–25, 262–263
Dirt, societal intolerance, 287
Disease(s)
  and bacteria, connection, 287
  and cheese, 281
  dengue fever, 110
  malaria, 80–81
  and microbes, 274, 281
  sickle cell anaemia, 81
  synthetic pathologies from *The Synthetic
    Kingdom*, 128
Disembodied Cuisine, 36
Disgust
  at human cheeses, 285–286
  and context, 290–291
Disposable products, 47
Disruptive technology, 103, 122–123
  synthetic biology as, *42*, 59
Distributed manufacturing, to reduce waste and
  energy consumption, 105, 154
Diurnisme (2007 Décosterd & Rahm), *64*
DIYbio, 59–60, 298, 300
DNA (deoxyribonucleic acid), 80–81
  atomic structure of single-stranded DNA, *81*
  automation of fabrication, 220

"backbone," 81

"bases," 82

  code, redesign, 54

  custom-writing, 21

  fall in cost of synthesis, 21–22

  first nonbiological message encoded, *32*

  "frame shifting," of triplet-coded DNA, 76–77

  images produced by gel electrophoresis, *240*

  original description, 9

  play and synthesis, 189–190

  possibility of sending digitally, 105

  as programming language, 11–12

  representation by gel electrophoresis, *240*

  representation by sound, *238–239*

  representation of molecule by letters, *234*

  synthesis, 22–23

DNA rooms, 224–230

  virtual, *228–229*

DNA sequencing, 75

  and bacteriophages, 75

  Sanger-based, 225, 227

  to identify bacteria in human cheese, 275, 278

DNA synthesizer, *22*

Dogs. *See also* Animals

  domesticated, and natural selection, 53

  ownership, as symbiotic interaction, 182

Do It Yourself biology movement (DIYbio),
  59–60, 298, 300

Domus (magazine), 130

Donkeys, and horses, to produce mules, 259

Dosier, Ginger Krieg (architect), *Better Brick*,
  122–123, *123*

Douglas, Mary (anthropologist), 288, 291–292

Drinking water, testing, 133

DTRA (Defense Threat Reduction Agency)
  (US), 60

"Dual-use dilemma," 60

Dunne & Raby, *Between Reality and the
  Impossible*, *Future Foragers*, 213

Dunne, Anthony (designer), 63–64, 67, 212

Dutton, Rachel, 282, 293

Dystopias

  in contemporary bio-fictions, 61

  dystopian extremes, 60

  "half-pipe of doom," 59

  and synthetic biology, 62

Earth's crust

  human influences, 196

  reformation, 200

E. chromi collaborative project, 68, *100*, 124–125.
  See also The Scatalog

  scenario set in 2039, 125

E. coli bacteria

  colonies expressing red and green fluorescent
    proteins in a biofilm, *152*

  long-term evolution, 119

  use in *E. chromi* collaborative projects, *xiii*,
    124, 131

Ecological crisis, 33, 47–48

Ecological diversity, and human cheese, 275, 278

Ecological footprint, in vitro meat, 36

Ecological harm, and synthetic biology, 115,
  130–131

Ecology, complexity, 204

E. coloroid bacteria, *xi*, 135–136, *135*

Economic implications of synthetic biology, xxi,
  66, 172, 180, 300

Ecovative Design, packaging from mycelium, 117

Ecuador, nature given constitutional rights, 130

Edinburgh. *See* University of Edinburgh

Educational models, biodesign in architecture
  studios, 154

E. glowli, *xiii*, 126

Electroencephalogram data from epilepsy, *236*

Electronics, analogies, and scaling down, 162

Elfick, Alistair, xviii, 28, 91, 94, 97

ELSI. *See* Ethical, Legal, Social Implications
  (ELSI)

Emmentaler cheese, *272*

  bacteria involved in production, 273

Endy, Drew, xviii, 59, 91, 94, 97

  Endy lab, bacterial data storage system, 119

Energy

  conversion, 121, 248

  new ways to use, 122–123

Engineering. *See also* Biological engineering

  and creativity, 184–185

  and design, xx

  evolution as challenge, 296

  history of profession, 185–186

  and invention, 187–188

  of nature, 190

  principles changed by biology, 23

  relationship with biology, 282

  use to redesign existing organisms, x

  vs. science, 262

Engineering design. *See also* Design

  and evolution, 266

  and synthetic biology, 97–98

Engineering logic, 28

  application, historical precedents, 29

  factory farming as application, 34, 209–210

Engineering mindset, 34

  and art, 37

  countering, 27–37

  and extremists, 34

  in life science laboratories, 33

Enterococcus faecalis bacteria, smell notes, *279*

Entropy, as useful design feature, 117

Environmental damage, caused by two centuries
  of industrial modernization, x–xi, *xiv*

Environment. *See* Biodiversity; Biology; Nature

Epilepsy, electroencephalogram data, *236*

Ethical design, definition of boundaries, 112
Ethical implications of synthetic biology, 42–43,
    46–47, 58, 61, 130, 172, 180, 298–300
Ethical, Legal, Social Implications (ELSI), 61.
    See also social, legal, political implications of
    synthetic biology
Ethics. *See* Bioethics
Eugenics, 31
Eureqa (software application), 147
Evaluation of new product types, criteria to
    use, 110
Everyday items
    as art or design, 212
    possibilities for synthetic biology, 172, *173*
Evolution. *See also* Directed evolution
    as algorithm to select optimized genetic
        designs, 16
    and biological design, 80, 113, 136, 255–256
    challenges, 296
    and change in living machines, 262
    of design, 100–137
    and design in nature, 78–79, 91, 93–94, 113,
        259–267
    and engineering design, 266
    and genetic mutation, 16
    and nature, xxi, 80, 82
    and obsolescence, 113–114
    optimization of genetic code to support, 82
    and progress, 53
    role in synthetic biology, 262–267
    as tyranny, 263, *264*
*Evolution of the Engineering Mindset Toward Life
    as a Raw Material*, 26
"Experiment," different meanings, 40
Extraterrestrial messages
    bacteriophage øX174, 77–78, *78*; news
        reports, 77
Extremists, and engineering mindset, 34

Factory farming, 34
FBI agents, at iGEM competition, 33
Federici, Fernan, 105, 108, 134–135, 296, 301.
    See also Bio Logic project
    bacteria, *xii*
Fermented drink. *See* Kombucha
Fiction. *See* Design fictions; Science fiction;
    Speculative fictions; Useful fictions
Films
    The Blob 1958, 60–61
    Soylent Green, 103
Flavor of cheese, source, 273
Flowers. *See also* Delphiniums; *Moondust*
    carnation; Plants; Venus flower basket sponge
    Common Flowers/Flower Commons, 129, *129*
Fly's eye, *79*
Food, and fuel, 107

Food safety, 287
"Fool's gold" (iron pyrite), *201*
"Foot" cheeses. *See* "Toe" and "foot" cheeses
Fossilized Stromatolite, *192*
Fraenkel-Conrat, Heinz, creation of artificial
    virus, 31
Franklin, Rosalind, 9
Fuel, for living and inanimate machines, 250
Fukuhara, Shiho, *Common Flowers/Flower
    Commons*, 129, *129*
Fuller, Buckminster, 65
Fungi, importance in cheese making, 272–273
Furedi, Frank (sociologist), 66
Fussenegger Group at ETH Zurich, 110, 111
    (caption)
Futility, engineering, 198
Future
    growing a biological future, 292–293
    responsibility of synthetic biology for, 209
    of synthetic biology, xiv–xv, 28, 40, 61,
        171–172, 205, 299–300

Gaia Corporation, and intellectual property, 129
Gamberti, Diego (sociologist), 34
Gas chromatography, of cheeses, *280*
GC-MS. *See* Gas chromatography; Mass
    spectrometry
"Gear changes," in sound composition, 223
Geitlerinema cyanobacterium, autofluorescent
    image, *198*
Gel electrophoresis, images, *240*
GeneForge Inc, DNA synthesizer, *22*
Genes, 10–11
    common features in all genes, 11
    nested, 75–77
    patentable, 130
Genetic algorithms, 109, 143
Genetically Engineered Sound Garden (Benque),
    66, *67*
Genetically modified organisms (GMOs),
    products synthesized by, 19
Genetic circuits, 18
    creation, 220–221
Genetic code, *83*, 265–267
    optimization in support of evolution, 82
Genetic diversity, loss of, 51
Genetic engineering, 19
    and synthetic biology, x, 20
Genetic modification (GM), 51
    new opportunities, 187–188
    public misapprehension, 61–62
    spread of variants, 107
Genetic mutation, as engine of evolution, 16
Genome
    "heredity element," 10–11
    of single cell, 10
Geological calendar, Western Australia, 200

Geological time
    and biological time, 198, 206
    and synthetic biology, 205–206
Geological timescale, *207*
Geometric shapes
    and quantifiability of human figure, *253*
    to construct ideal horse, 251, *251*
Ginsberg, Alexandra Daisy, 9, 33, 297–298. *See
    also E. chromi; The Synthetic Kingdom*
    (with Oron Catts) "The Well-Oiled Machine"
        (short story), 114
    (with Sascha Pohflepp), *Growth Assembly*, 104,
        104, 105, *105–106*
Glucose economy, 120, 153. *See also* Sugar
Google Inc. (search company), free time for
    employees to foster creativity, 189
Gorgonzola cheese, *272*
    fungi in production, 273
Government funding for synthetic biology, xiv
Greening Biotechnology, 38
Growth Assembly (Ginsberg and Pohflepp),
    103–104, *104*, 105, *105–106*, 107
Guthrie, Karen, *Cat Fancy Club*, *95*, 96

Haeckel, Ernst, 103
Hafnia alvei bacteria, smell notes, *279*
"Hand" cheeses
    bacteria isolated, *279*
    "Christina Hand" cheese, *278*
    odors, *279*
    smell notes, *279*
Harley-Davidson corporation, 88–89, 96
    engine sound, 89
    motorcycle undergoing sound testing, *90*
Haseloff, Jim, 101–103
Hearing, DNA constructions, 221, 237
Herbicide Gourd (*Growth Assembly*), *104–105*
Herbicide Sprayer (*Growth Assembly*), 105, *106*
Heritable information, 80
Hertog, Steffen (sociologist), 34
Hill Top Research Inc., Ohio, analysis of breath
    of research participants, *287*
Historical precedents, of engineering logic,
    application, 29
Historical progenitors of synthetic organisms, 209
History
    of engineering profession, 185–186
    importance, 214
    of the inanimate machine, 248
    of material appropriation by humankind, 7, 9
Holden, W.F.C., design proposal for Tower
    Bridge during WW2, *186*
Horsepower, 249–250
    to measure power of an engine, 13, *14*
Horses. *See also* Animals; Mules
    and donkeys, to produce mules, 259
    engineering design, 251

evolutionary advantage, 136
    as machines, 248–249
    as prototypes for "living machines," 260
Human beings, relationship with nature, 293
Human body. *See also* Bodies/the body; Body
    fluids; Body odor; Human microbiome
    biological link with cheese, 273
    Mushroom Death Suit, 117–118
    as site for manufacture, 293
    symbolism, 292
Human cheeses
    breast milk cheese, 290
    "Christina Hand" cheese, *278*
    "Christina Mouth" cheese, 285
    "Daisy Armpit" cheese, *278*, 285
    disgust at, 285–286
    from human bacteria, 275, *275–276*, 278, *278*,
        301
    microbes isolated from, *281*
    "Philosopher's Toe" cheese, *278*, 285
    "Sissel Nose" cheese, *278*, 285
Human era, effects on life and climate, 197
Human Genome Project, 61
Human intervention, mining and Earth's
    geology, *200*
Human microbiome, *289*, 293. *See also* Human
    body
    bacteria on cheese makers' hands, 283–284,
        284
Humility, and critical lens on synthetic biology,
    209, 214
Hygiene, nineteenth century movement, 286

"IDEAS Factory Sandpit on New Directions in
    Synthetic Biology," xvii, 299
IDEO, 136, 169, 178
    design process, *172*, *179*
    discussions with scientists, 180
    logo, xi, 135–136, *135*
    and science, *179*
iGEM competition. *See* International
    Genetically Engineered Machine (iGEM)
    competition
Imagining the future, effect, 68
Imperial College, London, in 2008 iGEM
    competition, 103
Industrialization, and synthetic biology, xiv, 107
Industrial Revolution, 12–13
    effects, 7, 9, 258
    and emergence of design, 44
    shaping of the 19th century, ix
Information Revolution, shaping of the 20th
    century, ix
Innovation
    or creativity, 184–186
    and play, 189–190

The Inside-Out Body project, xxii, 269–293
Instrumentalization
    of design and art, 69
    of life, 48, 114
Intellectual property
    for artists and designers, 130
    control by big companies, 128–129
    and *Gaia Corporation*, 129
Interdisciplinary collaborations, xvii–xviii, xx, 33,
    122, 133, 170, 282
    anthropology and synthetic biology, 133–134
    architecture and synthetic biology, 135, 155–164
    art and synthetic biology, 205, 210
    industrial design and synthetic biology,
        169–170, 180, 182
International Flavors and Fragrances (company),
    278
    gas chromatography-mass spectrometry of
        cheeses, *280*
International Genetically Engineered Machine
    (iGEM) competition, 13, 33, 154, 189–190. *See
    also E. chromi* collaborative project
    Arsenic Biosensor, 131
    Biofabricator, 103
    E. coloroid bacteria, *xi*, 135–136, *135*
    *Searching for the Ubiquitous Genetically
        Engineered Machine*, 131
    teams from 2012 Jamboree, *24*
International nature, iGEM teams, 2012, *24*
"Intolerable Beauty: Portraits of American Mass
    Consumption, Crushed Cars #2, Tacoma
    2004," *xv*
Introduction, ix–xxii
Intrusive technologies, long-term effects, 111
Invention, and engineering, 187–188
In vitro meat, 34–37. *See also* "Victimless" meat
    ecological footprint, 36
Iron pyrite ("fool's gold"), *201, 208*
Isovaleric acid, 274
    chemical structure of molecule, *273*
Iwasaki, Hideo, 70, 136, 297–299. *See also*
    Biogenic Timestamp project

Jacket, *Biocouture, 116*
Jacob, François (molecular biologist), 176–177,
    265
Japanese tsunami, and Biogenic Timestamp
    project, 203
J. Craig Venter Institute, 57, 263. *See also* Venter,
    Craig
Jeremijenko, Natalie (artist), *One Trees: An
    Information Environment*, 65
Jones, Horace, original design for Tower Bridge,
    185
Jonkers, Henk, *BioConcrete*, 118–119, *118*

Kac, Eduardo (bio artist), 69
    GFP Bunny, 69
Kay, Lily (molecular biologist and science
    historian), 252
Keratin cup, (*The Synthetic Kingdom*), *126*
Kill switches, 113–114
King, James, 103, 124. *See also E. chromi*
Knight, Tom, 12, 13, 263
Kombucha (fermented drink)
    food for, 120
    to make cellulose, 115, *116*, 117
Kornberg, Arthur, creation of virus infectivity,
    31
Kunsthaus Graz, 156, *157*

Lab Testing Rig (Benque), 67
Lactic acid, chemical structure of molecule, *273*
Lactobacillus bacteria, 272
    in yogurt, 287
Lactose, chemical structure of molecule, *273*
Lake Clifton, Western Australia, thrombolites,
    199, 201, 204
Land, use for food/fuel, 120–121
Language. *See also* Definitions
    of design, public awareness, 174
    different interpretations of "structure," 225
    different meanings attached to words by
        scientists and artists, xi, 40–41
Latour, Bruno (sociologist of science and
    anthropologist), 45, 64, 256
"Lease of life," 112–113
Leather
    from tissue culture, 120
    shagreen, *49*
    Victimless Leather, *35*
Le Corbusier (modernist architect), 57, *58*
Lee, Jae Rhim, *Infinity Burial Project*, 117–118
Lee, Suzanne. *See also* Kombucha
    BioCouture, 115, *116*, 117
Legal implications of synthetic biology, 24, 61
    "lease of life," 112–113
Leguia, Mariana, 136, 296. *See also* Synthetic
    Sound from Synthetic Biology project
Lenski, Richard, 256
    Long-Term Evolution Experiment, 119
Life
    and art, 202
    as biomatter, 28–29
    building from "bottom-up," 23, 54, 151, 254
    control by human beings, 195
    engineering, 28, 251, 256
    as frontier for exploration, 27
    instrumentalization of, 114
    materials, ownership, xi
    NASA definition, 262
    ownership, xi, 128–130
    as raw material, 28–29, 33

redesign, 48–52

representation and abstraction and, 242

tools to manipulate, 202

treatment by synthetic biology, 202

Life cycles, and life spans, 114–120

Life and death, as new design typology, 110, 120

Life spans

defining, 113

and life cycles, 114–120

Limburger cheese, odor, 273

Lim, Wendell, 135, 297, 299. *See also* Living among Living Things project

Lindbergh, Charles, 31

Lipid vesicles spontaneously forming structures morphologically similar to living cells, *254*

Living among Living Things project, xxi, 169–182

concept sketch, *173*

Living machines

mules as, 259–260

nature of, 53–54

Living Machines project, xxii, 247–258, 261–262, 267

Living rocks, *199*, *206*, *208*. *See also* Stromatolites; Thrombolites

Loeb, Jacob, artificial parthenogenesis, 29–30, *30*

Logic. *See* Engineering logic; Logic of biology

Logic of biology, 159–161. *See also* Bio logic project

and biocomputing, 150–161

London, Bernard (Manhattan real estate broker), *Ending the Depression Through Planned Obsolescence*, 112

Loss, and withdrawal, 241–242

Lumin Sensors, 131–133

Macfarlane, Robert (writer), 61, 66–67

Machines

living and inanimate, 248–250, 255

to mimic animal power, 249, *250*

Maize cultivars, development, *50*

Malaria, resistance to, 80–81

Mangrove

as example of nature, 93, 93 (caption)

roots and branches, *92*

Manifesto

for designers (*a/b*), 63, *65*

for liberation of engineers, 191

Mansy, Sheref, 136, 265, 296–299. *See also* Living Machines project

Manufacturing. *See* Biomanufacturing

Manzoni, Piero, *Merda d'artista*, 288

Margocsy, Daniel, 251

Mars, building materials, 153

Mass consumption, 44

Mass production, effects, 7, 9

Mass spectrometry, of cheeses, *280*

Materials. *See also* Building materials

abalone snail shell strength, 6

decomposable, 115–118

issues raised by new materials, 63

life's, ownership, xi

nature's use of raw material, and energy, 7

sustainable, 120

tree of materials, *8*

Mathematical models

to simulate complex biological behaviour, 144

xylem formation, 147, *148–149*

Matter, rearrangement though synthetic biology and art, 195–204

McDonough, William (ecological design activist), *Cradle To Cradle*, 115

McShane, Clay, 248

Meat

Churchill on growing meat, 34

Semi-Living Steak (TC&A), 69

Tissue Engineered Steak No.1, *36*

in vitro (*see* In vitro meat; "Victimless" meat)

Meinderstma, Christien, (designer), 50

Metal smelting, history, 7, 9

Metaphors, 163–164

DNA as dominant metaphor, 134

engineering, limitations, 296

metaphor and reality, 234

metaphor and scale, 162–164

in synthetic biology, 242

MICP. *See* Microbial-induced calcite precipitation

Microbacterium lactium bacteria, smell notes, *279*

Microbes

and cheese making, 274–275, 280

collected from human skin, 275, *275*

and disease, 274, 281

importance, 274

and improved health, 281–282

isolated from human cheeses, *281*

Microbial-induced calcite precipitation (MICP), 122–123

Microbial Kitchen, 67, *68*

Microbiolites, Western Australia, 201

Microbiome, human. *See* Human microbiome

Microscopy

confocal, and architecture, 158–159

as unfamiliar for designers, 170

Miller, Stanley, Miller-Urey experiment, 31, *32*

Mining, Australia, as example of human intervention, *200*

Modernism (20th century)

and improvement of the world through design, 57

negative effects, 58–59

Modern Meadow (research startup), 120

Molecular biology, as new science, 9–10

"Molting," and creativity in synthetic biology, 189

Monsanto, and "terminator" seeds, 114
Moondust carnation, 129
Moral issues around designing life, 59
Morganella morganii bacteria, smell notes, *279*
Mosquitoes
    and Limburger cheese, 273
    RIDL male (Oxitec Ltd), *109, 110*
"Mouth" cheese, "Christina Mouth" cheese, 285
Mules, 259–260, *260*
    as prototypes for "living machines," 260
Multicellular organisms, difficulty in applying
    synthetic biology, 23–24
Mushroom Death Suit, 117–118
Mushrooms, as building/packaging materials,
    117, *117*
Music. *See also* Acoustics; Computer music;
    Sonification; Sounds
    salsa music in different acoustic
        environments, 227, 230
    and structures, 220
Mycelium, for packaging, 117
Mycoplasma mycoides, synthetic ("Synthia"), *58*

Nanotechnology products, made from plant
    growth products, 4
National Aeronautics and Space Administration
    (NASA) (US), 262
National Bioeconomy Blueprint (US; 2012), 107
Nature. *See also* Biology
    constitutional rights in Ecuador, 130
    and design, 47, 90–91, 93, 180
    designed, 73–98
    as enemy, 61
    and evolution, xxi, 80, 82, 91, 93
    human beings and, 293
    influence on design and engineering, 143
    strength of abalone snail shell, 6
    and technology, 47–48
    unnatural, 52
    use of raw materials and energy, 7
    and values, 95
Nature is Designed, 72
Neeley, J. Paul (designer), *Gaia Corporation*, 129
Nepal, Practical Action (NGO), 133
Nepenthes bicalcarat (fanged pitcher plant),
    101–102, *102*
Niemeyer, Chris and Greg, *Tomato Quintet*, 227
"Night science," 177, 179
    and day science, 176–177
    and design, 178–179
    image, *166*
"Nose" cheeses
    bacteria isolated, *279*
    odors isolated, *279*
    "Sissel Nose" cheese, *278*, 285
    smell notes, *279*
Nozzle Fruits (*Growth Assembly*), 103–104, *105*

Obsolescence
    and evolution, 113–114
    and kill switches, 114
    planned, 112–114
Odors. *See* Smells
Olfactory illusions. *See* Smells
Open-Source Architecture, 130
Optimization, computational, 147–148, *148–149*,
    151, *153*
ORIs (origins of replication), 221
"Orthogonal systems," 54
Oshima, Tairo, 78
    original paper, *77*
Ownership, 59–60
    of dogs, as symbiotic interaction, 182
    of life's materials, xi
Oxitec Ltd, mosquitoes, RIDL male, *109, 110*

Packaging
    biodegradable and compostable, 117
    for biological consumer products, 132–133
    designing, 132–133
    that creates its contents, *171, 174, 176*
Parmesan cheese, *272*
    odor and olfactory illusions, 273
"Parts," sequences of DNA as, 232–233
Pasteurization of milk, 287
Pattern formation in xylem cells, 144
Paxson, Heather, 282
Pell, Richard (artist), 33, 52
    Center for Post-Natural History, 52
Penicillium fungi, and flavor of cheeses, 272–273
Perkin, William Henry, first synthetic dye, 48
Personal Microbial Culture, skin sample, *175*
Phage particles. *See also* Bacteriophages
    repurposing, 85
Philips Design Probe (*Microbial Kitchen*), 67
Philosophers of science, and use of models, 67
Philosophy, issues raised by new materials and
    experiences, 63
øX174 bacteriophage, *74*
    sequencing, 75
Phosphorus, 6
Photosynthesis
    cyanobacteria, 206
    discovery of, and art, 197
    effect on Earth's atmosphere, 197
"Physiological architecture," 63
Pig 05049, 50
"Pilbara craton," 200
Pinecone, as "seed dispersion subcomponent,"
    5, *5*
Pitcher plant (*Nepenthes bicalcarat*), 101–102, *102*
Plant growth products, nanotechnology
    products, 4

Plants. *See also* Acoustic Botany; Flowers; *Growth Assembly*
    designed, 50–52
    engineered by humans, 248
    fictional, 101–105
Plasmids, 220–221
    representation, *222*, 237
Plato, 184
Play
    and engineering, 187–188
    importance, 297–298
    and innovation, 189–190
    and synthetic biology, 189–191
Pohflepp, Sascha, 103, 136, 296–299, 301. *See also* Living Machines project
    (with Alexandra Daisy Ginsberg), *Growth Assembly*, 103–104, *104*, 105, *105–106*
Pohl and Kornbluth, *The Space Merchants* (1952), 34, 60
Poincaré, Henri, 184
Political implications of synthetic biology, xxi, 66, 107, 300
Pope, Nina, *Cat Fancy Club*, *95*, 96
PostNatural Organisms of the Month, *51*
Practical Action (NGO in Nepal), 133
Predator species, extinction as long-term effect, 111
Predictability, and biology, xi
Prejudices, about what we eat, 283–284
Presidential Commission for the Study of Bioethical Issues (US), 59
Probiotic bacteria
    for disease monitoring, 125
    drinks, 172, 173, 174, 176
    yogurt, 272, 274, 287
"Problem finders," designers as, 63
Product death. *See also* Obsolescence
    strategies for, 112–113
    and synthetic biology, 113–114
Product design, 115, 136
Products
    from biotechnology, 19–20
    synthesized by genetically modified organisms (GMOs), 19
Progress, and evolution, 53
Projects (Synthetic Aesthetic Residences)
    Biogenic Timestamp, xxii, 195–204, 208–214
    Bio Logic, xxi, 139–164
    The Inside-Out Body, xxii, 269–293
    Living among Living Things, xxi, 167–191
    Living Machines, xxii, 247–258, 261–262, 267
    Synthetic Sound from Synthetic Biology, xxi, 219–230, 232–233, 235, 237, *238–239*, 239, 241–242
Promoters, 14–15
Propionibacterium freudenreichii and flavor of cheese, 273

Proteus mirabilis/vulgaris bacteria, smell notes, *279*
Protocell researchers
    creation of living cells from scratch, 261
    and creativity, 261
    and engineering, 261
    synthetic creations as living things, 260–261
Protocells, 54
    creating, 23
Providencia mermicola bacteria, smell notes, *279*
Public awareness of language of design, 174

QWERTY keyboard, 189

Rabinow, Paul, 133
Raby, Fiona (designer), 63–64, 67, 212
Rahm, Philippe (architect), 63
    Diurnisme, 64
Registry of Standard Biological Parts, 130, 154
Reineck, Adam, 135, 136. *See also* Living among Living Things project
Replacement culture, 47
Replication, 255
    defined, 10
Representation, 232
    and abstraction, 231–242
    in art and science, 237–238
    loss produced by, 229–230, 235
Reproduction, engineered, fertilized eggs from cow's uterus, *111*
Responsibility
    for end-use of products, 47–48
    for future, and critical lens, 214
Restriction enzymes, 19
Reverberation, from synthetic biology, 220, 227
Revolution, design, 63
Rhode Island School of Design, 132–133
Ribosomes, 10
RIDL mosquitoes. *See* Mosquitoes, RIDL male
RNA (ribonucleic acid), 10–11
Rooms, different reverberations, 226
Ross, Phil, *Mycotecture*, 117, *117*

Safety, environmental, need for, *Moondust* flowers, 129
Salsa music, in different acoustic environments, 227, 230
"Sandpits," xvii–xviii, 299. *See also* "IDEAS Factory Sandpit on New Directions in Synthetic Biology"
Scale, and metaphor, 162–164
Scaling down, analogies with electronics, 162
The Scatalog, *100*, *124*, 125–126. *See also* E. *chromi* collaborative project
Schön, Erhard, print, *253*
Schwabe, Stefan
    The Kernels of Chimaera, 117

Living Artefacts, 117

Science. *See also* "Night science"; Synthetic biology
creativity in, 176–177, 179–180
design process to describe the scientific process, *179*
nature of, 233
positive effects of synthetic biology on scientists, 180
scientists and society, 62
vs. engineering, 262

Science fiction, 60. *See also* Wells, H.G.
as inspiration to scientists, 68
and potential hazards of biotechnology, 60
Soylent Green (film), 103
The Space Merchants (1952), 34, 60
and in vitro meat, 34–35
"The Well-Oiled Machine" (short story), 114

Science and technology studies (STS), 300
Searching for the Ubiquitous Genetically Engineered Machine (ArtScienceBangalore iGEM team 2011), 131, *132*

Sedimentary rocks, made by cyanobacteria, 199
Seeds, distribution, to replace actual products, 105

Self-contained production systems in synthetic biology, 115

Self-repairing materials, 119
Semi-Living Steak (TC&A), 69
Sensitivity of natural organisms, 6–7
Sequencing. *See* DNA sequencing
Shagreen leather, *49*
Shelley, Mary, *Frankenstein* (1818), 60
Shetty, Yashas, 131
Short-termism, and synthetic biology, 297
Sickle cell anaemia, 81
Silicon computer chip
digesting possibly to extract gold, 208
and sand, 85

Silicon dopants, 6
Silicon wafer containing computer chips, *4*
Simon, Herbert, 211
Simplifying biotechnology, 18
Simun, Miriam, breast milk cheese, 290
Skin-care products, designed for an individual, 174, *175*

Slime mold (*Physarum polycephalum*) growth, as solution to human problem, 161, *161*
Smelling breath of research participants, *287*
Smells
and cheese flavor, 273
differences between people and body parts, *279*
from different human cheese, *279*
olfactory illusions and cheese flavor, 273
smell notes of cheeses, *279*

Snow, C.P., 184, 282

Social implications of synthetic biology, xxi, 58, 61, 66, 135, 172, 180, 300
Social pollution, 288
Social scientists, involvement in synthetic biology, 300
Society, and scientists, 62
Sonification
definition, 219
diagram, *238–239*
everyday examples of, 235, 237
as form of representation, 238–239

Sonifying DNA, 216
Sound composition, description, 223–224
Sound reverberations, different rooms, 226
Sounds. *See also Acoustic Botany; Genetically Engineered Sound Garden*
derived from structure, 220
importance of right sound in Harley-Davidson engines, 89
synthetic, from synthetic biology, 219

Soylent Green (science fiction film), 103
The Space Merchants (novel; 1952), 34, 60
Specificity of natural organisms, 6–7
Speculative fictions, 66–68. *See also* The Scatalog
Speculative research, *132*
encouragement, 134
Spiegelman, Sol, "Spiegelman's monster," 31
Srishti School of Art, Design and Technology, 33, 131
Standards and standardization, 12–13
as engineering principle, 23
of measurement in biology, 14
standard-setting bodies, 13
and synthetic biology, 119–120
Steam engine, *250, 252, 255*
and evolution, 136
Steam/Metabolism, *244*
Steam production system, Watt, *252*
Steichen, Edward, Delphiniums exhibition, 96–97
Stromatolites, *185, 192, 204. See also* Living rocks; Sedimentary rocks
Structure
different interpretations, 225
and sonification, 220
STS (science and technology studies), 300
Sudjic, Deyan (design critic), 44
Sugar. *See also* Glucose economy
and biological products, 120, 254–255
Survival, importance in living machines, 255
Suspension of disbelief, and imagination, 56
Sustainability, 296–297
growth of sustainable materials, 120
Sustainable manufacture, 188
Symbiosis, 280
art and science, 282
designing, 130–133, 282

SymbioticA, 37, 204, 210
Symbiotic complexes, cheeses and bodies as, 280
Synthetic Aesthetics Residences, xviii, xx, 133–137, 301. *See also* Projects (Synthetic Aesthetic Residences)
    applications to participate, xviii
    artists' and designers' involvement with scientific tools and technologies, 301
    and DIYbio, 300
    as effective interdisciplinary collaborations, 300
    as fictions, 299
    and the future of synthetic biology, 298–299, 301
    and imagined future, 301
    origins, xviii, xx
    project team, xviii, xx
    as public engagement with science, 300–301
    questions raised by, xxi–xxii
    scientists'/engineers' involvement with art/design, 301
Synthetic biology, x, 3–25, 41. *See also* Biotechnology Revolution; Science
    alternative perspectives, xiv–xvi
    as applying engineering logic to life, 28
    beautification, danger of, 214
    collaborations between artists and engineers, 33
    conflict with art, 27–37
    and creativity, 188
    cultural significance, 197–198
    desire for standardization, 62
    in the developing world, 24–25
    dilemmas, 69
    as disruptive technology, *42*, 59
    dystopias connected with, 61
    as emerging interdisciplinary field, 197–198
    government funding, xiv
    potential, xiv–xvi, 39–40, 59, 61, 171–173, 197–198, 205, 299–300
    roles in our societies, 97
    technical ambition, x–xi
    "tilted hierarchy" of skills, 191
    as tool for mass production, 52
Synthetic chemistry, start of, 48–49
Synthetic Ecology, 131
The Synthetic Kingdom, *55*, *56*, *57*, 126, 128
    Carbon Monoxide Sensing Lung Tumour, *127*
    Keratin cup, *126*
    products, 128
Synthetic life. *See* Life, as raw material
Synthetic Sound from Synthetic Biology project, 219–230
"Systems," and creativity, 188

Tarr, Joel, 248
Tatum, Edward, "one gene, one enzyme" hypothesis, 75
Technologically Mediated Victimless Utopia, 35–36
Technology, and design, 187–188
Thackara, John (design innovator), 46
Thacker, Eugene, 19
Three-dimensional computer modeling, 147
    printed prototypes, *150*
Thrombolites., *199*, 201, 204, *208*. *See also* Living rocks; Sedimentary rocks
Timber, as building material instead of steel, 115
Time
    as critique, 205–214
    and life, 209
Time-dependency, of cyanobacteria, 199
Timescales, and cyanobacteria, 206
Tissue culture
    development of technique, 30–31
    DIY techniques, development, 128–129
    leather from, 120
    "semi-living" status, 69
Tissue Culture and Art Project (TC&A), 35–36, *35–36*, 69, 204
    semi-living steak project, 69
Tissue Engineered Steak No.1, *36*
"Toe" and "foot" cheeses
    bacteria isolated, *279*
    collecting bacteria, *275*
    odors isolated, *279*
    "Philosopher's Toe" cheese, *278*, 285
    smell notes, *279*
Tolaas, Sissel, 136, 296–297, 301. *See also* The Inside-Out Body project
    smell molecule studio, *xix*
Tools, first use, 7
"Top-down." *See also* "Bottom-up" systems
    control of computer chip technology, 4
    design, and modernism, 57, *58*
    engineered bacteria, 54
    synthetic biology systems, 254
Tower Bridge, London, evolution of designs, *185–186*
Transcription, 10
Transcriptional promoter, 11, 14
Transcriptional terminator, 11
Translation, 10
Translation start and stop codons, 11
Tree of fossilized life, *8*
Tree of life, *8*. *See also* The Synthetic Kingdom
    as human construct, 53
Tree of materials, *8*
Tremmell, George, *Common Flowers/Flower Commons*, 129, *129*
"Triplet codes," 76, 78
Tumors, *Carbon Monoxide Sensing Lung Tumour*, *127*

Turing patterns, 156–157, *158*
"Two cultures," 184, 282

United States
    Defense Advanced Research Projects Agency
        (DARPA), 60
    Defense Threat Reduction Agency (DTRA),
        60
    National Aeronautics and Space
        Administration (NASA), 262
    Presidential Commission for the Study of
        Bioethical Issues, 59
    White House *National Bioeconomy Blueprint*
        (2012), 107
University of California at San Francisco
    (UCSF), 180
University of Edinburgh
    2006 iGEM team, 131
    biosensor development, 133
Unpredictability, 163
"Up-skilling," 18, 25, 188
Urban development, Western Australia, threat to
    thrombolites, 201
Urey, Harold, Miller-Urey experiment, 31, *32*
Useful fictions, 66–68, 137
Utility, and uselessness, 208, 210–211, 297–298
"Utopia point," 109, 112

Value(s), 96–97
    in design, importance, 300
    and nature, 96–97
    role in design, 87, 89–90, 97, 298–299
    and synthetic biology, 96
    of work that does not have an obvious utility,
        208
Venter, Craig, 57. *See also* J. Craig Venter
    Institute
    "first life form whose parent is a computer,"
        31, 33
    "synthetic cell" creation, 261
Venus flower basket sponge, *84*
Vesicles spontaneously forming structures
    morphologically similar to living cells, *254*
Victimless Leather, *35*
"Victimless" meat, 36, 69. *See also* In vitro meat
La Ville Radieuse model, *58*
Virus, artificial, creation, 31
"Voice labeling," 223

Waseda University of Tokyo, 210
Waste, 115
    possible synthetic biology solution, 3–4
Water. *See* Drinking water
Watson, James, 9
Watt, James, 13
    steam engine, 252, 255

steam production system, *252*
"The Well-Oiled Machine" (short story), 114
Wells, H.G., 27
    The Island of Dr. Moreau, 29, 60
    "The Limits of Individual Plasticity," 27, 29
Western Australia, 200
Wiener, Norbert, *Cybernetics, or Control and*
    Communication in the Animal and Machine,
    251–252
Williams, Reid, 135, 325. *See also* Living among
    Living Things project
Williams, Robley C., creation of artificial virus, 31
Wittenberg, Jacob (undergraduate computer
    music student), 227, 230
Worms, *Caenorhabditis elegans*, 131
Wyndham, John, *The Day of the Triffids*, 60

Xylem
    digital models of xylem cells, 159–161
    exoskeletons, as biodesign tool, 144–151
    formation in epidermal cells of *Arabidopsis*
        thaliana, *146–147*
    structures in nonvascular cells, 146

Yeadon, Peter, 132–133
Yeast
    chassis organism, 16
    engineering to produce commercial products,
        254
    in production of cellulose, 115
Yokoo, Hiromitsu, 78
    original paper, *77*

Zurr, Ionat, 43, 325